SAFETY 24/7

BUILDING AN INCIDENT-FREE CULTURE

Gregory M. Anderson &
Robert L. Lorber, Ph.D.

SAFETY 24/7

BUILDING AN INCIDENT-FREE CULTURE

Greg Anderson
PO Box 99823
Seattle, WA 98139-0823
832.654.9099

Printed in the United States of America
ISBN: 0-9778308-0-2

Credits
Design, art direction and production   Back Porch Creative, Plano, TX   info@BackPorchCreative.com

FOREWORD

Why would anybody write a whole book on safety?

The reality is, every organization has a "safety culture"– whether good or bad. What all organizations need is a *culture of safety*, where everyone takes personal responsibility for their own safety as well as those around them.

Regardless of the industry you work in, or what country you may live, no one ever wants to get hurt on the job – or any other time, for that matter. To create an environment where individuals are motivated to work safely, be personally accountable for their actions and willing to provide feedback to their teammates requires a commitment at all levels of an organization.

It's interesting to note, while Walt Disney declared his theme park would be the "happiest place on earth," he designated safety as the number-one value. His picture of perfection was that guests would leave the park wearing the same smile they had when they arrived hours earlier. As a result, he created a culture of safety while establishing an environment of courtesy and friendliness that all of us have grown to know and love.

Whether you are in the energy sector, mining, engineering and construction, or just care about the people around you, *Safety 24/7* provides the tools necessary for creating the foundation of a great organization, no matter what your mission may be.

After reading this book, I am a raving fan of the work Bob Lorber and Greg Anderson have done. Bob has been a friend and colleague of mine for over 25 years. As coauthor of *Putting the One Minute Manager to Work*, he made a major impact by helping millions of people put the three secrets of the *One Minute Manager* into practice. I have known Greg since he was a teenager. His father, Gordon Anderson, is one of the most creative and innovative managers I have ever known. Greg has obviously inherited those good genes. This book is the culmination of Greg's experience and Bob's fine mentoring.

Thanks, Greg and Bob, for focusing our attention on this vital aspect of running a great organization – and on the need for each of us to be aware of the importance of *Safety 24/7*.

Ken Blanchard
Co-author of *The One Minute Manager*®
San Diego, California
January 2006

CONTENTS

INTRODUCTION

Will you be going home from work today? What about the person next to you?

On average, 16 fathers, mothers, sons and daughters won't make it home from work each day. Why? Because in the United States more than 5,000 workplace fatalities and 1,500,000 lost time injuries occur each year. What's more, according to the United Nations' International Labour Organization, over two million lives are lost each year to job-related accidents and illnesses, plus more than 268 million lost time accidents occur globally.

Disturbing statistics? They disturbed us enough to write *Safety 24/7* to show you how these incidents can be dramatically reduced, even eliminated. *Safety 24/7* is also about how each of us impacts safety and how our individual commitment will make a real difference in everyone's life.

Nobody wants a fatality to occur or an injury to happen. Organizations try to eliminate these incidents by making safety their number one priority or writing additional safety policies and procedures ... but they fail, time and time again. Why? Because they have left the human element out of the equation.

Safety 24/7 introduces you to a proven process that will help eliminate incidents, not only in the workplace but in your everyday life. It isn't a difficult book to read. Unfortunately, many of its lessons have been tested by those who can no longer tell their stories.

We hope you will use *Safety 24/7* in your personal journey and, as a result, will work to continuously improve safety wherever life takes you. Remember, real change can only take place where there is an open mind and a willing heart.

THE CHALLENGE BEGINS

"Manager of worldwide safety."

Kurt Bradshaw liked the sound of his new position and repeated it as he dialed the phone to share the news with his wife, Jessica.

"I'm so proud of you, Kurt. This calls for a celebration."

"We can also celebrate the raise that comes with the title."

"More money? Even better."

"I'm on the way to a meeting with my new boss now, but I'll see you at 6:00."

Sitting in senior vice president of operations Ron Kaiser's office, Kurt's new responsibilities became instantly more challenging as the man across the desk ticked off his expectations.

"We've got to turn things around," Kaiser was saying. "The new policies and procedures we put into place six months ago aren't working. We had two fatalities last month in South America, four Lost Time Accidents in West Africa and who knows what we never hear about. Our claims cost … not to mention our lost productivity … is driving our stock down.

"Bottom line, Bradshaw, we've tried everything and nothing seems to work, including the zero tolerance program we initiated last year. So, we need change and we need it fast. We think you're the man for the job.

"You can start with South America," Kaiser said, pointing to the map behind his desk. "Their numbers are high, so it will be our test site for whatever changes you decide to make … oh, and we want results in 120 days. Then, with any luck at all, we won't have the accidents we're having now."

"That deadline doesn't give me much time," Kurt countered.

"That's true, but our reputation in the industry is starting to suffer. We need results." Kaiser's tone let the new manager of safety know there was no room for discussion.

Back in his office, and after dealing with a steady stream of well wishers, Kurt decided to call John Sullivan, his co-worker and friend in South America.

John and Kurt had become close over the past few years, close enough for their families to spend time together last year when John came to the States for the company's annual meeting.

A native Australian, John had begun his career at the bottom of the ladder and had progressed upward, much the same way Kurt had done.

"Couldn't have picked a better man, my friend." John sounded genuinely happy to hear Kurt's news. "We'll still be working together. That's the main thing."

"But worldwide safety … it's a huge job … and our numbers aren't impressive," Kurt responded. "Bill Andrews worked hard, putting new policies and procedures in place, but our incident severity and frequency rates continue to rise, which – no doubt – was the reason they made a change … and Kaiser made it clear, he wants results, so I may be the proverbial lamb on my way to the slaughter."

"Never let 'em see you sweat, mate. Besides, you're up to the challenge."

John's words echoed as Kurt opened the door after dinner that evening. "After you, ladies."

"Don't think I'll eat again until Thursday," Jessica declared.

"Me, too … but I'm going up now … still have homework." His daughter Shannon stood on tiptoe to kiss her dad. "Thanks for dinner, you're the best," she whispered before dancing up the stairs.

At 2 a.m., Kurt had yet to close his eyes, thinking of the job ahead of him. "I know operations. I've proven that," he thought to himself, "but safety … what do I know about safety? It's an entirely different world than operations."

Besides all the policies and procedures in place, the company has invested millions to upgrade the equipment – with little improvement. What about the program, "Safety is our Number One Priority?" Everyone seemed fired up, but then the momentum tapered off, safety disappeared from individual radar screens and the number of fatalities and injuries kept climbing. What is missing …?

The next morning at breakfast, Jessica poured their second cups of coffee and brushed Kurt's cheek with a kiss. "You look tired."

"Didn't sleep much. Kept thinking about Kaiser's deadline and trying to remember as much as I could about what has been done in the past. Jess, I'm not sure I'm the right guy for this job," Kurt admitted. "I don't have a background in safety and I have no idea how to stop all the accidents. What do I really bring to the table?"

"Kurt Bradshaw! Are you saying you aren't the best man for the job?"

"No. It's just that I've had a chance to think about what Kaiser expects – and it's a tall order … maybe darn near impossible," Kurt said. "You know the old phrase, 'Over-commit and under-deliver?' I'm just wondering if I've over-committed by saying I would take the job."

Kurt's wife listened in silence.

"Just look at what they're expecting."

"But remember how much operations improved when you became manager? Your team would follow you anywhere. How were you able to do that?" asked Jessica.

"I always worked with good people," Kurt countered.

"Your leadership skills, your ability to talk with anyone on the team and provide feedback, your obvious caring for each person and helping them make the most of their opportunities," his wife continued. "That's what set you apart. That's why management thought you'd be a good manager of safety."

Kurt finished his toast and the last swallows of his coffee as he remembered some of his past achievements.

Now it was Jessica's turn to sip her coffee thoughtfully before she looked up. "It may be you're just being modest," she suggested, "but, if you're feeling unsure about the whole proposition, what about some help? Remember Dad's old friend, Sam Rollins?"

Kurt remembered the name. Sam Rollins had been honored at an industry seminar. "Yeah, I saw him get a safety award a while back. Really impressive background. He's been around long enough to know this industry inside and out."

"Dad thinks so highly of him. He might be a good sounding board."

"At least he might give me some ideas about where to start," Kurt's mood brightened at the idea. "I'll call him and see if he has time to meet."

As Kurt kissed his wife goodbye, Jessica tucked a slip of paper into his coat pocket. "While you were dressing, I called Dad and got Sam Rollins' number."

"What would I do without you?" Kurt asked between hugs. "See you tonight."

"Good luck, Mr. Safety Manager," Jessica called behind him.

"Sam Rollins," a deep, gravely voice answered Kurt's call.

"Mr. Rollins, my name is Kurt Bradshaw. My father-in-law, Hal Rankin, gave me your number."

Explaining the turn of events leading to his call, Kurt was pleased Sam was available the next day.

"Why don't you come to my house tomorrow evening at 6:00pm," Sam suggested.

✦ ✦ ✦ ✦ ✦

"Ten minutes ahead of schedule." Kurt turned off the car and picking up his computer bag, walked across the front porch and pressed the doorbell.

Sam answered the door with a wide smile and a generous handshake. "Welcome, Kurt. I'm glad you took me up on my offer to meet here. Come in, come in. Make yourself at home and I'll get us some coffee."

The foyer was impressive, decorated with plaques and several shelves of photos. Kurt noticed one photo of Sam standing among what appeared to be a crew on a drilling rig. Another showed Sam and two people who could be his wife and son. The last photo was a much younger Sam in a military uniform, standing with several other soldiers.

In a few minutes, Sam reappeared, carrying two mugs of coffee. "Follow me. The front porch is comfortable this time of day."

After the usual pleasantries, Kurt got right to the point of his visit. "Sam, I've been in this business for over 10 years, but this is the first time I've had the deadlines and responsibilities for safety I have now," he began. "To be honest, I'm looking for some help on where to begin."

Sam stroked his mustache before he answered, his eyes focusing on an invisible scene in the distance. Then his gaze returned to Kurt. "There's no doubt. You have a big job ahead of you, but it isn't an impossible task.

"In a nutshell, your job is to create a strong safety culture for your organization, which means a lifestyle and belief system for everyone who works there now and in the future, because a *culture* is made up of group behaviors and beliefs that are transmitted from one generation to the next," Sam explained.

"One of the biggest challenges you'll have in improving the culture that already exists is dealing with human nature and *old school* behavior – but, as a manager, I'm sure you've already had a taste of that."

Kurt agreed, eager to hear more.

"In your business – in any business – you'll find there are things people do, day in and day out, that put them or someone else at risk. I call these *at-risk behaviors*, and they are the root cause of most incidents. They are also the foundation for what I call the safety pyramid," Sam explained.

Fatality

Lost Time Injury

Recordable

Near Hit

At-Risk Behaviors

Looking at the pyramid Sam had drawn, Kurt asked, "Why do you say *incident* instead of *accident*, and don't most organizations use the term *near miss* instead of *near hit*?"

"You're very observant," Sam replied. "But, no, I use the word *incident* because *accident* implies something happened outside of someone's control, which is not the case 97 percent of the time. And, when you think about it, what's the difference between a near miss and a near hit? So why do we call it a near *miss*? Typically we use words like *accident* and *near miss* to lessen accountability or minimize the potential consequence. That's literally like whistling by the graveyard," Sam sighed.

"Getting back to that old school mindset – like having to prove how tough you are to make it in this industry – perpetuates many of these at-risk behaviors. Remember old school isn't about age. Today's workforce is made up of as many young dinosaurs as it has old ones. An organization's culture creates them."

Kurt nodded his understanding. "It's easy for a new-hire to pick up that attitude. In no time at all, you can't tell the newcomers from the old timers. They all work the same. Mind if I write a few things down while we're talking?"

"Be my guest," said Sam, pausing as Kurt took a pad of paper out of his bag.

"Another mindset is what I call *the bullet-proof* mentality," Sam continued.

"Bullet-proof folks – particularly young people, but don't exclude older workers from this group – think they know it all, particularly their jobs … and they don't think they'll ever be hurt. It's that tendency to think, 'Nothing can hurt me because I'm a young invincible stud' that gets them injured or killed. As they get older, that idea becomes, 'Nothing's hurt me yet, so why should I worry?'

"Take wearing safety glasses, for instance. What do you *expect* to happen if you don't wear them," Sam asked.

Kurt thought for a moment. "Well, usually nothing except you might get yelled at … or at least that's my experience."

"Exactly!" Sam said, leaning back in his chair. "But, what if – while you were working – something broke and sharp pieces went everywhere or, worse yet, hit you in the face?"

"I'd make sure to wear my safety glasses in the future."

"The at-risk behaviors we take when we're in an old school or bullet-proof mindset – like not wearing those glasses – set the stage for many potential outcomes." Sam explained, drumming the table with the knuckle of a missing forefinger. "You could be lucky enough to never be hit in the eye or you could be unfortunate enough to be blind for life, or even killed. The only place we have any control over the outcome is at the beginning, at the bottom of the pyramid, when we make the choice whether or not to wear our safety glasses.

"Here's another question for you," Sam probed. "Do you believe people in your industry can do their jobs with zero incidents?"

"Well … I'd like to hope so … but it's a high-risk industry and we've come to expect accidents … I mean incidents … will occur," Kurt responded hesitantly.

"The issue of *what we expect* is very important in establishing a strong safety culture," Sam explained. "Years ago in Los Angeles, a new teacher came into a school and was told she would be teaching a class of the school's brightest eleven- and twelve-year-olds. The principal told her the sky was the limit for these kids.

"At the end of the school year, achievement tests showed this new teacher's class had improved their scores more than any other class in the school. Most of the students had been so hungry to learn, the new teacher had to supplement the existing curriculum.

"Now here's the kicker. That new teacher was actually given some of the poorest performers in sixth grade. But, because she expected them to excel, they did."

"So, how did she manage to improve their performance?" Kurt asked.

"It was all a matter of what the teacher thought about the students – and her expectations," Sam said. "Remember, the principal told her the sky was the limit for her class. That's how the teacher saw the students, and superior performance was what she expected."

"Expectations had no limit," Kurt commented as he made more Notes. "On the other hand, if the principal had told the teacher she was getting a class of poor performers, she might have had low expectations. I know my crew had the highest production rate when they knew what my expectations were. I guess the same is true for safety. People will work to the level of safety expected of them."

"Exactly," Sam smiled. "You're a quick study … we're going to get along fine."

Kurt looked across the table, quizzically.

"Oh, don't think this is going to be our last discussion. In fact, I'd like to meet with you on a regular schedule, to discuss your progress and help you develop the process necessary to establish a strong culture of safety," Sam offered. "I say *process* rather than *program* because a program implies it will have an end, and safety is a never-ending effort."

"I'd be glad to meet regularly. Over the next 120 days, I'm going to need all the help I can get," Kurt smiled wryly.

"Deal," said Sam, extending his hand. "Looking forward to it."

Driving back to the office, Kurt felt a relief of sorts. "Sam knows his stuff," he said to himself. "On the other hand, with the timeframe I've got to work with, Sam needs to help me come up with a way to change our culture fast," Kurt thought as he parked his car. "and with operations running day and night, not to mention our shortage of personnel, we need a safety process people will use both at work and at home, 24/7."

Snapshot

Definitions:

1. **At-risk behaviors:** those actions we take day in and day out that put us or someone else at unnecessary risk.

2. **Incident vs. accident:** an accident implies the result is outside our control. In 97 percent of the cases, what happens – the incident – is easily within someone's control.

3. **Near hit vs. near miss:** maintains focus and accountability on the potential consequence.

4. **Old school mindset:** creates an environment that gets in the way of speaking our concerns.

5. **Bullet-proof mentality:** thinking we won't be hurt as a result of our actions or behaviors.

Essential Concepts:

✦ Our expectations are at the core of building a culture of safety and improving performance.

✦ People will work to the level of safety that's expected.

Fatality
Lost Time Injury
Recordable
Near Hit
At-Risk Behaviors

✦ The Safety Pyramid illustrates how an at-risk behavior can easily escalate to become a Lost Time Injury or even a fatality. The only place we have control over the outcome is at the base of the pyramid when we choose to do, or allow, an at-risk behavior.

2   NEW IDEAS, OLD ISSUES

Kurt looked forward to his Friday afternoon meeting with Sam. "I don't know what I was expecting," Kurt admitted easing into a rocking chair on Sam's porch. "I attended safety meetings at several locations, which didn't offer much in the way of surprises. But, I also consciously observed what people were doing and almost immediately saw at-risk behaviors, from personnel not holding hand rails nor using safety equipment to people leaning back to put their feet on the table while they balanced on two legs of their chair and …."

"I imagine you observed more than you wanted to," Sam finished Kurt's sentence.

"More than I expected … so I need your help in sorting it all out," Kurt nodded. "I started by asking people how safe they felt on the job – and their answers were across the board. One guy – probably in his early 20s – said he knew people got hurt, even killed, at work but didn't think it would happen to him … or even if he broke an arm or a leg, it would be no big deal."

Sam nodded knowingly. "What that guy has yet to figure out *is everything we do – from the moment we're born – carries an element of risk … and his survival depends on how well he manages those risks.*"

Kurt continued, "Another worker, who looked as though he had plenty of experience, said as long as he did the job the way he had always done it and nobody had gotten hurt, he didn't think there was any risk.

"Then a couple of people said they thought about safety on the job because they had to. They cited programs like *Safety is Our Number One Priority* and admitted they thought it was a joke, but complied with it because they were told to and didn't want to lose their jobs. Sure enough, when the project fell behind schedule, deadlines became a priority and the focus on safety disappeared.

"Another guy said he practiced safety measures at home as well as on the job … it was a way of life for him. But, here's something interesting. He said he talked with his wife about using a cutting board as a safer way to cut vegetables instead of cutting them in her hand; but when it came to work, he admitted he was hesitant to talk to his team-mates about their at-risk behaviors.

"Out of all the people I spoke with – probably 20 or more – I only found two who actually seemed passionate about safety. One was a fairly new employee. The other had been working for about 15 years and both said they were willing to shut the job down when they observed at-risk behaviors, even if their supervisors were pushing them to meet a deadline.

"When I asked why they were so motivated to work safely, one of them said his dad was an electrician and taught the whole family to be safety-conscious, so it was second nature to him. The other employee said he learned the hard way … he saw one of his buddies killed in an incident that could have been prevented. After that, he opted to go with his gut about doing the job safely … and he didn't care what anybody else thought.

"So that's it in a nutshell. Lots of attitudes out there, huh?" Kurt turned to a clean sheet on his note pad as he finished reading through his list.

Sam put down his mug and leaned back. "Good work … and very impressive, too. You haven't wasted any time discovering ***a person's attitude toward safety is a choice***, and there are four levels of commitment to safety you'll find on every job and in every person. If I can borrow your pad, I'll draw them out for you:

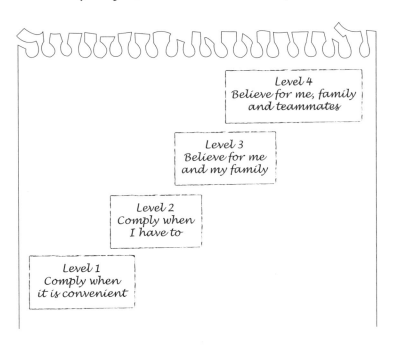

Kurt looked closely at what Sam drew. Then he noticed something. "I think I get it. Level 1 and 2 say, *comply* – meaning I do it because I'm told, forced, or paid to do it. At Levels 3 and 4, the attitude changes to *believe* – meaning I make a conscious choice to practice safety and am totally committed to working in that mindset. And when I reach Level 4, I'm demonstrating a strong culture of safety."

"That's right," Sam interjected. "But I want to add one reminder, each of us is a blend of all four levels of commitment ... so just because a worker demonstrates Level 4 most of the time, you need to expect he may revert to Level 1 when he doesn't know the potential danger."

Kurt nodded. "I'm glad you mentioned that ... so, Level 4 behaviors aren't constant?"

"Exactly – and here's another twist," Sam said, putting down his mug to make the point. "Level 2 behaviors – complying with safety measures because we have to – will often look like Level 3 or Level 4.

"Does a person wear his personal protective equipment because the boss is around or because he truly believes in safety for himself and his teammates? Ask yourself ... why do you put your seat belt on when you drive? Do you do it because the police will stop you if you don't?"

"Well, I ... er ...." It was a question Kurt had to think about. "I'm trying to find the honest answer," he explained. "I've been using my seat belt so long, it's just a habit. When I first started, I used it because I knew it was the law, but when I read about people being killed and they weren't using their seat belts, it fueled my Level 4 mindset."

"Here's another one," Sam grinned, "and these are not trick questions. Do you slow down when you see a police car sitting on the side of the road ... or do you normally observe the speed limit?" He paused.

"My point is this ... you have to look beyond actions to determine the attitude causing the behavior. Those different attitudes impact the safety culture an organization builds.

"Do you remember how we defined *culture* the last time we talked?" Sam asked.

"Isn't culture made up of group behaviors and beliefs that are transmitted from one generation to another?" Kurt answered.

"Exactly," Sam said. "Now I want to expand on that further. ***We create our culture by what we demonstrate personally and by what we reward and tolerate in others.*** Think about what you've seen your organization reward and tolerate."

Kurt was silent for several seconds before he responded. "I'm sure we must be rewarding some of the right behaviors … but, on the other hand, we're also missing a lot of the at-risk behaviors.

"For example, while I was in the break room, a person tossed a canned soft drink to a co-worker. At the same time, someone stood up to leave the room and almost got hit by the flying soda can.

"As another example, I was standing in the elevator when the door began to close. A guy stuck his arm between the doors, just so he wouldn't have to wait for the next one. He was lucky enough to squeeze into the elevator before it started to move, but he could have lost fingers, a hand … or worse."

"And what about you – how did you demonstrate a personal commitment to safety?" Sam asked.

"Well, I spoke up when I observed at-risk behaviors. When the guy pried his way onto the elevator, I asked him if he knew the risks of what he had done … but before he could answer, a woman standing next to me said she had just read about a doctor at a hospital being beheaded when he got stuck in the door of an elevator that began moving.

"I probably also surprised a lot of people in the break room. After I saw the guy toss the soda, I pointed out the risk involved. Although the guy was embarrassed, he hadn't considered the possibility of his behavior causing injury," Kurt explained. "I reminded him he'd been lucky this time, but also explained the importance of safety even during breaks."

Sam applauded. "That's a giant step, Kurt … thanks for having the courage to speak up. If you'd taken the easy way out and stayed quiet, you would have perpetuated a culture that tolerates at-risk behaviors. Your safety conversations demonstrate your Level 4 commitment to safety. In the strong safety culture you want to build, safe behaviors become a habit, *a way of life*, for everyone because they're automatically reinforced.

"Have you or Jessica ever tried going on a diet?" Sam questioned.

Kurt chuckled, "You mean our annual New Year's resolution that lasts about 2 weeks?"

"About 10 years ago, my weight shot up to almost 300 pounds." Sam explained. "I tried a few diets without success and then asked my doctor for some of those diet pills you hear about. Well, he refused, saying what I needed to do was make healthy eating a way of life. Following a diet isn't a way of life. Neither is taking pills. The doctor told me I needed to relearn how to eat … so I could get back to the 225 I'd weighed for years."

Kurt had trouble even visualizing Sam at 300 pounds.

"The doctor helped me make simple changes to the way I ate and exercised … but changes I could live with on an on-going basis. Needless to say, I didn't drop a whole lot of weight real fast. It took months, but I learned there's a real difference between being on a short-term diet versus adopting healthy eating as a lifestyle. Once you make it a habit, you can keep it going over the long-term."

"Okaayyy," Kurt said slowly. "So what does that have to do with establishing a safety culture?"

"Putting a new policy or procedure in place after an incident is just like taking diet pills after gaining 75 pounds," Sam explained. "Both help us feel we are doing something to solve the problem but neither creates sustainable long-term results. Remember the guy who talked about *Safety is Our Number One Priority*? He said the organization emphasized that program until the project fell behind schedule and then the customer's timeline replaced safety as a priority.

Sam shook his head as he continued. "The cold, hard facts are, just like dieting usually doesn't result in permanent weight loss, policies alone cannot guarantee a totally safe workplace … no matter how hard we try."

"So, *a strong safety culture begins when we start making safety a habit?*" Kurt asked.

"A habit you practice every minute of the day," Sam pointed out. "Yes, that's a good first step and, speaking of steps, do you know what a step-change is?

"Only if you're talking about country western dancing …" Kurt laughed.

"No, Tex, I'm talking about a step-change in safety," Sam said with a smile. "The oil and gas industry improved its safety performance in identifiable steps. Before the 1960's, we were in the *Era of Death*. Individuals in our industry were viewed as hard-working men slinging dangerous steel and we simply accepted *accidents will happen.*

"Our first real step in improving safety came though adding things like handrails, limit switches and guards on machinery. *Companies* expected fewer people would get injured as a result of these technological improvements. This *Era of Engineering* did reduce injuries and fatalities, but over time safety plateaued as people became complacent with the new equipment and the opportunities to engineer safety became fewer.

"Once our industry realized technology and engineering didn't prevent incidents, we took the next step … the *Era of Legislation*. New laws were put into effect and organizations wrote hundreds of policies. *Government* expected fewer people would get injured as a result of increased regulation. Unfortunately, it didn't prevent one of our industry's worst disasters from taking place. Ever see that before?"

Kurt studied the photograph of an offshore platform engulfed in flames and billowing black smoke. "Is that Piper Alpha?"

Sam nodded. "1988 … killed 167 men and changed the lives of hundreds of co-workers and families who lost friends and loved ones."

"I can't imagine what it would do to Jessica and Shannon if I were killed," Kurt sighed.

Sam continued, "Subsequent investigation, detailed in the Cullen Enquiry, found *mechanical processes and procedures alone do not create a safety culture. A lack of following procedures, a failed permit to work system and an unsinkable attitude among workers* all contributed to the tragedy. Despite the best equipment and all those policies and procedures, safety still came down to people."

"Did any good come out of everything that happened?" Kurt wanted to know.

"Absolutely … it ushered in the *Era of Behavior-based Safety*. Don't get me wrong, the work is still hard and dangerous and, unfortunately, incidents still happen, but now it is the *individual* who expects not to get injured … you see, **creating a safe work environment is a 'personal' issue, as well as a corporate one.**"

"I remember the last time we met, you said 'We get the level of safety we expect,'" Kurt said thoughtfully.

"You've got a good memory," Sam complimented. "During the Era of Engineering, it was the company who expected to improve safety …"

Kurt completed the thought. "And in the Era of Legislation, governments expected it, but the critical step in improving safety results depends on individuals expecting zero incidents."

"What we've learned is technology and policies alone don't eliminate incidents. Take the military – a huge organization - for example," Sam continued. "Have you heard how much emphasis they're putting on reducing personnel injuries and incidents?"

"Sure, I just read an article that despite all the state-of-the-art equipment and lots of regulations, incident rates are near an all time high. But look how dangerous their environment is," Kurt replied.

"No one would argue the risks associated with being in a *hostile* environment," Sam agreed, "but the majority of incidents taking place are non-combat related. The military is taking steps to reduce these incidents through implementing new regulations and technologies, but just like the oil industry evolved from the *Era of Death* to the *Era of Behavior-based Safety*, reducing at-risk behavior and eliminating incidents requires a step-change in culture where individuals expect not to get injured."

"Well, Sam, you've given me a lot to think about ... and you've helped me make some sense of what I've been seeing and hearing over the past two weeks." Kurt paused to chuckle. "Talk about a fast track ... I'm going as fast as I can, but I still have a long way to go. That 120-day deadline is looming large."

Sam nodded his approval, then reached over to tap Kurt's head. "They've got it here, son. Now, you've got to get them to feel it here," said the older man, tapping his chest.

SNAPSHOT

ESSENTIAL CONCEPTS:

✦ Everything we do carries an element of risk … and survival depends on how well we manage those elements of risk.

✦ Each of us chooses our attitude toward safety.

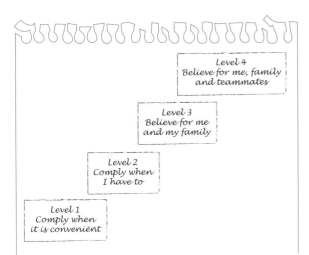

✦ Levels of commitment to safety: Everyone is a blend of these various levels. The question becomes "which level of commitment do we demonstrate most frequently?"

✦ We create our culture by what we demonstrate personally and by what we reward and tolerate in others.

✦ A strong safety culture begins when we start making safety a habit.

✦ Creating a safe work environment is a *personal* issue, as well as a corporate one.

ACCENTUATING THE POSITIVE – AS EASY AS ABC

As Kurt continued to visit the company's various locations, he began to realize establishing a strong culture of safety for the entire organization would be the test of his management career.

To create a true safety culture, poor safety habits would have to be identified and differentiated from proper practices. In most cases, that would mean changing individual attitudes. He decided the first step was to develop a process that would enable employees to make a personal commitment to work safely.

It was clear to Kurt, his organization – like many – had spent a lot of time and money initiating numerous safety programs. However, very little time had been spent on follow-up and reinforcement.

"How can you accomplish measurable results when you spend most of your time implementing a new program?" he wondered as he went through the large stack of folders he had brought home that evening.

Long before his alarm went off the next morning, Kurt's phone rang.

Fumbling for the receiver, Kurt managed a hoarse hello.

"Kurt, it's Tom Montano in Venezuela. Sorry to wake you, but I'm afraid I have some bad news."

Kurt sat up.

"John Sullivan's been injured in an automobile accident," Montano continued. "He's in the hospital. I don't know how bad his injuries are. They're trying to stabilize him now."

Now Kurt was wide awake. "I'll get down there as soon as I can," he said. "How's Theresa doing?"

Hearing her friend's name, Jessica sat up to listen.

"Oh, she's pretty shook up," came the reply. "We all are."

"Tell her I'm on my way and Jessica and I have both of them in our thoughts."

"Will do," Montano said. "Let me know your arrival time and I'll send a driver."

"Sure thing," Kurt said, hanging up the phone.

"How bad is it?" Jessica asked gently.

"Don't know, but bad enough for John to be in the hospital after a car wreck," Kurt replied. "I'll have to get a flight out as soon as I can."

"Why don't you jump in the shower while I start laying out some clothes for you to take," Jessica offered.

"Not before I have a hug," her husband took her into his arms. "I love you so much … and I'm praying John is going to be okay."

"Poor Theresa," Jessica whispered in her husband's ear. "John is her rock."

"I know. I know," Kurt answered.

It was after nine that morning when Kurt finalized arrangements for his flight to Venezuela.

"Kurt," Sam's familiar voice greeted him. "Just wanted to call about our meeting this week."

"Sam! I'm glad you called. I'm on my way to the airport. There's been a car crash in Venezuela and I need to get down there … but hey, how are you at spur of the moment trips? I think your safety expertise could be a real plus on this one."

"I keep my passport updated," Sam said, "I could be packed and at the airport in an hour."

"Let me clear it upstairs."

After reaching altitude, the drone of the jet's engine settled into a steady rhythm. Putting down his newspaper, Kurt turned to Sam. "I've got to confess, I'm really worried about John. He's a good man and valuable to the company, but more than that, his wife and son rely on John to make their worlds go round." Kurt remembered the times he had spent with John and grew quiet.

"Before I left the office, I found out a few more things about the incident and as I listened to the details, I found myself looking at the situation from a different viewpoint," he continued.

Sam sat quietly, allowing Kurt to talk as much as he wanted about his friend and the crash.

"Three people from our Caracas office had spent the evening at a client reception. During the party, their driver became too ill to drive. Apparently, the guys were in a hurry to get back to their apartments, so they decided to drive themselves.

"Montano said John was thrown from the car because he wasn't wearing a seatbelt … which is unbelievable. John is such a stickler for safety. He's not only that way at work, he's that way at home and with his family.

"Perhaps I'm trying to over analyze the crash because I'm beginning to recognize people's at-risk behaviors."

"That's definitely a good first step," Sam said. "Your perspective and awareness are definitely changing how you look at other people. Remember though, for real change to occur, behavior has to change, yours included."

"What do you mean?" Kurt asked.

"You've probably flown on airplanes a lot in your career – so much so that you unconsciously put on your seatbelt when you sit down. But let me ask you, where are the life jackets located?" Sam said watching Kurt closely.

"Underneath my seat, and my seat cushion can be used as a flotation device," Kurt said, mimicking a flight attendant.

"On this airplane, the life jackets are between the seats," Sam explained. "Imagine trying to figure that out in a smoke-filled cabin."

"I guess I should have listened to the flight attendant a little closer. Boy, do I feel like a Level I," Kurt said, putting his newspaper in the seat pocket in front of him. "And I'm supposed to help our company create a strong safety culture?"

Sam chuckled. "You are definitely not a lost cause, but in order to maintain a clear vision of safety, you have to mentally practice identifying your own at-risk behaviors.

"Before we talk about the next step in building a culture of safety, I want to show you some photographs." The older man unzipped his bag and retrieved a small photo album. "These photos are several years old, but they make a point."

As he looked through the album, Kurt saw a toddler he estimated was about a year old. In the first shot, she was standing with Sam who was holding her hands above her head as she took a clumsy step. In the second, Sam had let go of her hands.

The third shot showed the little girl taking a step on her own. Successive shots showed her trying again and then falling into waiting arms. In the final shot, she was sitting in Sam's lap, enjoying a cookie.

"She's cute," Kurt said, handing back the album. "I'm assuming she's your granddaughter."

"Yes." Sam nodded, smiling. "But these aren't just proud grandpa shots. I brought them along because they illustrate something you need to know."

"Okay," said Kurt, feeling as though he may have missed the obvious.

"It's pretty basic, really," Sam said, replacing the album into his bag. "Those photographs illustrate the *ABC's of performance*, a key point in changing behaviors. In the first photograph, I stood Erika up and held her hands over her head to keep her from falling. I call that the **Activator** – what I needed to do before I could expect Erika to take her first step – and to give her confidence she could do it.

"In the next picture, you saw her take a step by herself – that's the **Behavior** I wanted. But, then she fell down and had to be coaxed a bit to get up and try again, which she did. In the final shot, you saw the **Consequence** – the positive feedback she received from me, a hug and a cookie."

Kurt smiled, remembering the similar process he and Jessica had used to encourage Shannon to learn to walk.

"Of course, building a strong safety culture isn't as simple as offering a person a cookie if they do what you ask them to do, but the principle is the same: **a**ctivator-**b**ehavior-**c**onsequence," Sam continued.

"I didn't realize it before, but I used the ABC's every day when I was in operations," Kurt recalled. "It started by making sure my team had clear goals. Then I checked on them during the job to see how they were progressing and, finally, I always followed up to be sure they clearly understood the result of their efforts.

Kurt looked out the window, as he thought about previous jobs he had done. "I also found my team liked to be *caught doing something right*, which was a concept I learned at a seminar several years ago. *Accentuating the positive* seemed to motivate everyone on the crew, regardless of how long they had been with the company."

"That's because responding to positive reinforcement – like Erika and the cookie – is a human characteristic," Sam replied. "To give you an example, several months after my granddaughter was born, her big brother Kevin was having a hard time going from being the only child and the center of attention to being Erika's big brother.

"It didn't take him long to figure out if he couldn't get attention one way, he would get it another – like writing on the walls or putting marbles down the toilet. He didn't care if the attention received was negative. All that mattered was the focus was on him.

"My daughter quickly decided to teach him the ABC's at an early age. She started to redirect, or **activate**, Kevin into another activity, like coloring on a piece of paper, which changed his **behavior** and resulted in a positive **consequence**."

Again, Kurt thought about what he and Sam had discussed. "I now understand how important the ABC's are in creating the type of culture we need to reduce our incident rates."

Sam smiled at his young friend's progress. "Yes, but keep in mind while the ABC's will help change behavior, it must be clear those changes will result in the culture you are trying to create … and let me tell you what I mean by that.

"As you know, there is a change room on drilling rigs where personnel put on their work clothes just before going outside. On one rig, there were several instances when guys stepped outside the change room and, because they didn't look around, walked underneath loads that were being moved by the crane. One person even got hit in the head.

"To fix the problem, a light was placed above the door that led outside the change room. The light would start blinking when the crane was in operation, reminding workers to look up for the moving load when they walked outside.

"One day, though, a guy saw the light wasn't blinking ... so he stepped outside and got hit in the head by a load being moved. On that particular day, as it turned out, the light bulb had burned out.

"Unfortunately, having everyone focus on the light as they went out created the wrong behavior, which resulted in the same consequence they were trying to avoid."

"I know what you mean about the solution sometimes creating the behavior you are trying to change," Kurt nodded. "Because I drive a company car, I had to attend a defensive driving course. I was taught things like how to get out of a spin, backing-up quickly and emergency braking. When I got out of the course, I thought I was even more invincible."

"You're right," replied Sam. "Teaching someone how to get out of a spin without first addressing the choices they made which put them in that position is like closing the gate after the horse is out. It's great if a person knows how to handle a dangerous situation, but I'd ask whether it was necessary to put themselves at risk in the first place?"

Kurt nodded his agreement, "Just like the blinking light on the rig."

"Ladies and gentlemen." The pilot's voice interrupted from the aircraft's intercom. "We'll be landing in Caracas in about an hour."

"This afternoon has gone by more quickly than I anticipated," Kurt said, taking in a deep breath and letting it out slowly. "When we land, I want to go directly to the hospital. I need to find out how John – and his family – are doing. I'll have to admit, this whole thing has torn me up."

"Incidents and injuries affect many more than just those who were physically hurt. It's like dropping a pebble in the water and watching the circles spread across the entire pond," Sam said almost in a whisper.

"Your being here for John and his family is a good thing. It will motivate him to get better, and motivate us to make sure it doesn't happen to anyone else in the company."

SNAPSHOT

ESSENTIAL CONCEPTS:

- ✦ For real change to occur, behavior has to change ... ours included.

- ✦ To maintain a clear vision of safety, each of us has to mentally practice identifying our own at-risk behaviors.

- ✦ The ABC's of Performance: **A**ctivator-**B**ehavior-**C**onsequence. Clearly communicate the goal you expect, follow-up to see how they are progressing and at the end, be sure they see the result of their efforts.

- ✦ Noticing when people do something right – or "accentuating the positive" – is a strong motivator because responding to positive reinforcement is a human characteristic.

- ✦ To change a behavior, redirect the person to a different goal and provide a positive consequence when he or she achieves it.

- ✦ Incidents and injuries affect many more people than just those who were physically hurt – much like the ever-widening ripples caused by dropping a pebble in the water.

HUMAN NATURE IN A HIGH RISK ENVIRONMENT

4

It had been unseasonably warm in Caracas. As usual, the daily rains brought by the Wet Season continued to come hard and fast before dissipating quickly. However, the warmer temperatures had created foggy conditions in the valley, making visibility particularly treacherous late in the evening.

John had been driving the night of the crash. After attending a client reception, he and two co-workers were driving home when John swerved to miss something on the road. The car had spun out of control, causing the vehicle to leave the roadway and roll into a deep gully.

Tom Montano provided some of the details as they left the airport, heading for the hospital in the city's center. "John's in pretty bad shape ... two broken legs," Tom began, "but the doctors believe he is out of danger. The first few hours after the crash were touch and go because of internal bleeding, but surgery seems to have corrected those problems."

"That's good to hear," Kurt said, obviously relieved.

"As an expatriate myself, I'm surprised John was driving," Sam commented, knowing local drivers usually drove ex-pats in most of their international operations.

"Apparently the driver had been ill all day and, during the course of the evening, his stomach pains grew worse," Montano said. "You know how John is … he always wants to help. He released the driver for the night and planned to call another driver. When no one was available, John offered to drive."

"But, why was he the only one thrown from the vehicle?" Kurt wanted to know.

By now, they had pulled into a parking space near a gleaming multi-storied building. Montano grimaced before answering. "As I mentioned on the phone, I don't think he was wearing a seatbelt. The others were."

Kurt shook his head in disbelief. John was one of the most conscientious, law-abiding men he knew. Surely he hadn't overlooked something as simple as fastening a safety belt.

"All right if we leave our stuff in the car?" Sam wondered out loud as he stood next to the vehicle.

"The driver will be here," Montano said. "We can't stay long. Visiting hours are over in 15 minutes."

Theresa saw the trio coming down the long hall and met them halfway. "John's been sleeping most of the day," she said, giving Kurt a big hug as she struggled to maintain her composure. "But he'll be so happy to see you."

Kurt took a moment to introduce Theresa to Sam, who excused himself to the waiting room around the corner with Montano.

As Kurt followed Theresa to her husband's bedside, he remembered her as a beautiful woman with unforgettable eyes. Today her face was lined with worry and her eyes dulled by lack of sleep.

John's room was dark, except for a dim light over the hospital bed where lines and pulleys kept John's legs elevated in their heavy casts. His face was swollen, scraped and cut in several places, including a large gash, now sutured, that traveled well into his hairline.

"John, Kurt is here," Theresa whispered.

The man in the hospital bed opened his eyes slowly. "Hiya, Mate," his Australian accent clearly evident in spite of his condition. "Did a pretty good job of getting you down here, eh?"

"What are you doing in bed? You're not due a vacation for months," Kurt joked.

"Some vacation," John said, attempting to smile.

"I hear you're going to make it," Kurt reassured, reaching to touch his friend on the arm. "Keep up the good work ... and try not to give your nurses a hard time."

The man in the bed nodded and closed his eyes, dozing from the drugs he had been given for pain.

"Is Ethan here?" Kurt asked, as he and Theresa stepped back into the hall.

"I told him to stay at school," she replied. "He's getting ready for first-quarter exams and, besides, what can he do?"

"He could be here to support his mother ... and his dad," Kurt answered, knowing Theresa's independence wouldn't allow her son to miss classes.

"I'll be fine," Theresa said. "We talk every few hours, so he's keeping up with what's going on. I've hired a private nurse to keep a close eye on John, so I think I'll go home tonight ... sleep in my own bed."

Kurt gave her another hug. "I'm staying at the Melia Caracas," he said, holding the woman at arm's length. "I expect you to let me know if you need anything."

"Thank you for being here, Kurt," she said, squeezing his arm before returning to her husband's bedside.

✦ ✦ ✦ ✦ ✦

Later that night after Kurt had spent time with the men riding in the car the night of the wreck, he and Sam sat down for a late dinner.

"I've known John for years," Kurt was saying. "I've worked with him and I just cannot believe he took the risk of getting behind the wheel without being that familiar with the road ... or the automobile he was driving. It's so out of character for him."

Sam put down his menu. "Could be his natural tolerance of risk coming into play."

"What do you mean?" asked Kurt, trying to decide between camarones and pescado.

"We naturally tolerate some risks because we don't even notice they exist – we aren't aware something could be dangerous. In John's case, had he ever been on that particular road before? Was he aware of the potholes or the winding stretch where the accident occurred?

"In other situations, we become complacent because we've done something so many times without a problem, we take it for granted we won't have problems the next time we do the same thing. John may have thought, 'I've got a good driving record. I'm a safe driver so I'm the best one to handle this.'

"Finally, there are some people – probably not John – but there are folks who actually enjoy the thrill of taking risks. They get a 'rush' when they take chances … like speeding through heavy traffic, changing lanes and weaving around other cars."

"So risk tolerance should always be factored into the equation when you're looking at an incident and its causes?" Kurt asked, closing his menu.

"That's right … ***most injuries and incidents are the result of people's natural risk tolerance.*** But more importantly, before starting, or anytime there is a change during the task, a person should factor in their risk tolerance by asking themselves three questions," Sam emphasized as he began writing on a napkin.

1. Am I aware of all the risks associated with this task?

2. Have I become complacent as a result of doing this task many times before?

3. Have I allowed my desire for a thrill to impact the way I will conduct this task?

"Montano said something about rushing to get home before midnight, because of a conference call scheduled for early the following morning. That still doesn't tell me why John would take the risks," Kurt rubbed the fatigue from his eyes, "or why the other guys wouldn't try to stop him?'"

"Those are good questions," Sam said, "but I want to try and help you get at the basics here."

"Okay, shoot."

"First of all, you need to realize thousands of incident investigations – like the one you're doing now – are conducted everyday … *after* someone is hurt. Now that John is in the hospital, we look at all the things that should have been considered earlier."

A waiter interrupted Sam to take their orders.

"Do I hear you saying people get hurt because they don't make conscious choices how not to get hurt?" Kurt continued, once the waiter had left the table.

"That's correct, Sam said. "You may assess the risks when you decide, 'I'm not going to get hurt' … but you still consciously need to think through, *how* am I going to stay safe, despite the potential risks? It's that conscious thought – that conversation, whether with yourself or your team – that's key."

"And that conversation can take place anywhere at anytime, 24/7," Kurt replied, as the waiter reappeared with their dinner. "Now let's enjoy this feast before we both fall asleep."

✦ ✦ ✦ ✦ ✦

Before going to bed Kurt made a call to Montano, asking if he could see the wrecked automobile. Then he called Sam's room.

"I'm going to go and look at the car first thing tomorrow," he said. "Want to go?"

"If it will work into your schedule, why not just drop me at the office on your way," Sam suggested. "I'd like to spend some time talking to the people there."

"Sounds like a plan," Kurt responded. "Montano will pick us up at seven o'clock. I'll see you in the morning."

The car appeared to be a total loss as Kurt walked around the white sedan. One tire was blown out, windows were smashed and the frame was badly bent. Then something caught his eye. The driver's seatbelt was buckled and the left portion of the belt had been ripped from its bolted position on the floor.

A wave of relief flooded Kurt's thoughts. "I knew John wouldn't drive without a seatbelt," he said to himself, "and that tire … could it have blown, causing the car to go out of control?"

But what remained of the wrecked vehicle still didn't answer his question about why John decided to risk driving that night.

Sam's words from the night before returned. "We have to consciously recognize the hazards."

"But, isn't it possible that everything around us is a hazard?" he said aloud.

"What?" Montano asked.

"Oh, sorry. When you're thinking about doing things safely, would it be possible for everything around us to be a hazard?" Kurt repeated.

"Hadn't thought about it like that before," Montano replied. "It's possible I guess ... like that rock over there?"

"Right. ***A hazard is anything that has the potential to contribute to an incident taking place***, so that rock is a hazard," Kurt explained. "But because hazards exist all around us, we have to continually risk assess to prioritize and focus on those hazards that have a greater likelihood of contributing to an incident, as well as its severity."

"Let me put it this way – as long as the rock is over there and I'm over here, the likelihood the rock will contribute to an incident is very low. But, if that same rock were close to me – maybe a portion of it is hidden beneath the car – the likelihood of me kicking it or tripping over it increases significantly."

"Now that we're talking about it, I probably wouldn't have thought about the rock, or anything else for that matter, being a hazard," Montano said.

"Maybe not consciously, but our five senses – sight, hearing, touch, smell, and taste – are continually processing the environment around us for potential danger. We need to mentally practice safe behavior by raising it to a conscious level," Kurt continued, "so with that in mind, we have a two-step thought process to use every time we assess risks," Kurt pointed out.

"The first step is to ask ourselves, 'What has the *potential* to cause an incident?' And remember, there are hazards we don't even recognize as potential dangers because we are unaware, have become complacent, or are looking for a thrill, also known as our natural risk tolerance.

"The second step in the process is to ask ourselves, 'What is the *likelihood* the hazard will contribute to an incident?' But, just like the first step, my natural risk tolerance may cause me to underestimate the hazard."

Montano took a moment before recapping. "So, ***managing our safety starts by consciously recognizing the hazards*** and then determining the likelihood of an incident occurring, in one of three ways:

(1) Remote – very unlikely it will happen

(2) Possible – there is a chance of it happening but it's not inevitable

(3) Probable - likely or inevitable

"You've got it," Kurt said. "Now, as I look at this wreck, something else occurs to me that I'll share with you. ***Risk assessment is driven by the task, not by the people doing the task.***"

Montano looked puzzled.

"Let me explain it this way," Kurt offered. "Let's say you're an experienced driver, so you may feel the task of driving is not a risk to you because you know how to drive. But, my point is, risk is based on the potential hazards associated with the task itself, not the risk relative to you. Your experience helps reduce the hazards associated with the task – in this case driving – and lessons the likelihood and severity of an incident taking place.

"In the case of John's crash, I think the incident that night might have happened no matter who was driving," Kurt went on.

"Visibility, familiarity with the road, or the potential of a tire blowing out are a few of the hazards that existed, regardless of who was driving. Problem was, because of his natural risk tolerance, John didn't do anything to reduce any of the hazards that existed.

"What he should have done was think, 'What task – driving – am I about to do? What are the potential hazards? What's the likelihood of an incident occurring? How severe could the incident be if it were to occur? How can I reduce those hazards that exist?'"

"What if John was unsure about any of the answers?" Montano asked.

"If he was unsure, or something in his gut told him something wasn't quite right, he shouldn't have gotten behind the wheel," Kurt responded.

✦ ✦ ✦ ✦ ✦

Back at the office, Kurt briefed Sam on his findings.

"Well, you certainly have had a productive morning," Sam said.

Kurt was ready to get right to his questions. "If a person's natural risk tolerance, or at-risk behavior, impacts their ability to do a task safely, who is responsible for calling it to their attention? Take John as an example. He was so eager to get the job done – in this case get home and get some sleep – that he didn't think about the risks associated with driving; however, there were two other people in the car …."

"Remember, Kurt," Sam interrupted. "In a strong safety culture, everyone is responsible for speaking up when they observe a person performing an at-risk behavior, regardless of the reasons why."

"I realize that, Sam. What I meant was, there were two other people in the car who could have said something, but why didn't they?" Kurt replied, finishing his sentence.

"Often times the *culture* prevents people from speaking up because they don't know what to say or how to say it."

"What do you mean by the word *culture*," Kurt wanted to know.

"Well, in many cases a national employee may feel uncomfortable speaking up to an expatriate, which is a *cultural* barrier. Other times a subordinate may have difficulty speaking up to a supervisor, which speaks to an organization's culture. In either case, the key is to create a level of trust that allows people to speak up. Everyone, especially supervisors, has to learn **when it comes to safety, disagreement does not mean disrespect**."

"Talk about a culture change …" Kurt said, slowly getting up to leave for the day.

 # Snapshot

Definitions:

1. **Natural tolerance for risk:** Individuals naturally tolerate some risks because they don't even notice they exist – they aren't aware something could be dangerous.

2. **Complacency:** We become complacent because we've done something so many times without a problem. We take it for granted we won't have problems the next time we do the same thing.

3. **Thrill seekers:** People who actually enjoy the thrill of taking risks. They get a *rush* when they take chances … like speeding through heavy traffic.

4. **Hazard:** Anything that has the potential to contribute to an incident taking place.

Essential Concepts:

✦ Most injuries and incidents are the result of people's natural risk tolerance.

✦ Before starting a task, we can reduce our risk tolerance by answering these three questions:
   1. Am I aware of all the risks associated with this task?
   2. Have I become complacent as a result of doing this task many times before?
   3. Have I allowed my desire for a thrill to impact the way I will conduct this task?

✦ People get hurt because they don't make conscious choices how not to get hurt.

✦ In addition to deciding "I'm not going to get hurt," we still consciously need to think through, *how* am I going to stay safe, despite the potential risks? Whether talking to ourselves or our team, this conversation is the key to our safety.

✦ Managing our safety starts by consciously recognizing the hazards … and realizing everything around us has the potential to contribute to an incident.

✦ Risk assessment is driven by the task, not by the people doing the task.

✦ In a strong safety culture, everyone is accountable for speaking up when they observe a person performing an at-risk behavior.

✦ A *culture* may prevent someone from speaking up because they don't know what to say or how to say it." Therefore, a culture of safety creates a level of trust that encourages people to speak up.

✦ When it comes to safety, disagreement does not mean disrespect.

The Basis of the "Safety 24/7" Conversation

The early morning breeze revitalized Kurt as he walked up the hospital steps.

His daily visits seemed to help speed John's recovery. As Kurt entered the room, John hoisted himself with help from the trapeze hanging overhead.

"You still here?"

"I'm a bit of a captive audience, wouldn't you say?" John shot back.

"Well, I didn't come to see your ugly face anyway. Where's Theresa?"

"Oh, she's talking to the doctor about my physical therapy that begins this afternoon," John announced, unable to hide his excitement. "The doctor let me know it'll be a long road to recovery. From here, I'll go to a rehabilitation center for a month, but I may make it out of there before the holidays ...."

Kurt didn't want to dampen John's spirit, but his job required having this conversation with his friend, sooner or later. "So, how do you feel about discussing the wreck?"

"I'll do my best, Mate. What do you want to know?" John asked haltingly.

"I've looked over the incident report, but I want you to tell me what happened," Kurt said, looking through his papers.

Straightening his back, John began. "About 10:30 p.m., I decided it was time to head home, but my driver got sick earlier in the evening and I released him to go home, assuming I could get a hold of another driver.

"But when I couldn't, and realized I was the only one who hadn't had anything stronger than lemonade to drink that night, I decided to drive. The other two hopped in and promptly fell asleep.

"We had been traveling about 10 minutes and visibility was decreasing because of the fog, so I slowed down and literally crept around the turns. All of a sudden something appeared ahead of me – maybe an animal – I don't know. I veered to the right, trying to avoid whatever it was and ended up going off the road. I was trying to do the right thing … I …."

Kurt looked squarely at his friend. "It's not a matter of right or wrong. The issue is how could you have gotten home safely?

"I realize the circumstances changed … you didn't go to the party planning to drive home … but you know it's against policy for an ex-pat to drive in Venezuela," Kurt continued. "Another issue is the condition of the vehicle. Did you know anything about the car before getting behind the wheel?"

"Like what? A late model sedan … I think it had a half-tank of gas, but that's about all," John recounted.

"When I went out to look at the car, I found the seatbelt had ripped from its mooring," Kurt said. "Overall, the seatbelt was in pretty bad shape, just waiting for the impact of a crash to send the driver flying."

John was silent for a moment. "I'll admit, my judgment was not the best," he finally managed, "but none of us could come up with a faster way to get home, so I was behind the wheel by default … and – please understand, Mate, I'm not making excuses."

"So tell me, how much time did you save?" Kurt paused to let the question hang in the air. "I know you're a good driver, and probably didn't think there would be any risks, but you also didn't decide how you *weren't* going to get hurt."

"So, how does that work?" John asked. "Of course hindsight is 20/20, but it would be good to know for future reference."

Kurt was pleased with John's response. "I'm glad you asked that question. As I learn more about developing a culture of safety, I'm finding we need to place more emphasis on assessing the risks of everything we do, whether it's something like driving, how we sit at our desks, or go up and down stairs."

"It's a simple process, really," Kurt went on. "Before we start a task, like you driving that night, we need to ask ourselves three questions:

1. **What am I doing?** I'm driving a car back to Caracas.

2. **What might be hazardous about what I'm doing?** I'm not familiar with this vehicle or road. Fog is limiting visibility and humidity is creating a slick roadway.

3. **What can I do to reduce those hazards?** And, your best answer would have been, "We need to call another driver to come out and pick us up, someone who knows this vehicle, the road, and knows how to drive in fog."

"The point is this: *Risk assessing is consciously and purposefully deciding how we are not going to get hurt.*"

"So, is it possible to eliminate all risk?" John asked.

"Unfortunately, not," Kurt responded. "But, you bring up a good point. *Risk is determined by the probability of an incident occurring, given the associated hazards*, and a hazard is anything that has the potential to contribute to an incident taking place. Therefore, the level of risk a person faces depends on the presence or absence of hazards. To put it simply, *the more hazards we can eliminate, the lower the risk.*

"It does sound pretty simple," John said. "And something anyone can ... I mean should do."

"Well, I'd better be going. I have work to do and people to see," Kurt said as he gave his friend's arm a reassuring squeeze. "I'll be leaving tomorrow, but I'll stop in before I go."

✦ ✦ ✦ ✦ ✦

After lunch, Kurt and Sam were on their way to Lake Maracaibo to visit a barge and attend safety meetings. They asked Tom Montano to come along in case they needed an interpreter.

When they arrived, Guillermo, the barge master, welcomed them warmly and began the tour. As they walked, Sam noticed a man atop an eight-foot ladder, painting the crew's living quarters ... but what caught his eye was the lack of proper fall protection. Instead of using a real harness, he was tied-off by what looked like an old piece of rope.

"Guillermo," Sam said. "Do you mind if I try to talk with one of your people for a minute?"

"Sure … is there a problem?" Guillermo asked looking around.

"Just a small one, but I want to make sure it doesn't become a bigger one," Sam said walking over to the painter. "Hi, can I talk with you for a few minutes?" Sam asked, to see if the man spoke English.

"Yes sir," said the worker, untying his rope and climbing down the ladder.

Sam introduced himself and Montano. "I was observing you paint, and I'm real pleased to see you have your safety glasses on. But there's something that concerns me about your safety. Do you know what it might be?"

"The ladder is pretty old I guess …" the painter said, looking around to see who was watching.

"What would happen if that old ladder broke while you were up there?" Sam explored a little further.

"I could fall … but that's why I tied myself off," came the reply.

"And I appreciate your taking the time to do so," Sam said, genuinely. "Do you normally use that rope to tie off?"

"Not all the time," the painter said, holding up the piece of rope, "I usually use whatever I find laying around the paint locker."

Being careful not to sound judgmental Sam asked, "What's the worst thing that could happen if you fell and the rope broke?"

"I guess I could hurt myself," the man quietly replied.

"What would happen if you fell on your head?" Sam pushed a little harder.

"Well, I guess I could break my neck … but I'm real careful and have never fallen before," the painter answered quickly.

Seeming to change subjects, Sam asked, "Do you have a wife and kids at home?"

"Yes, a wife and four children … three boys and a girl."

"You must be very proud," Sam smiled. "What would happen to them if you were hurt and couldn't work anymore?" Sam asked, making sure the conversation stayed on track.

"It would be really bad, because we take care of my wife's parents too," came the serious reply.

"Is there something better you could use to tie-off with?" Sam asked.

"Well, I know the drilling crew use real harnesses. Maybe I can borrow one from them."

"That would be great! Then you wouldn't have to worry about not being able to take care of that wonderful family of yours. Can I get your agreement you will use a real harness from now on when working on a ladder, or more than six feet off the ground for that matter?" Sam said reaching out to shake the man's hand.

"Okay … sure I can," the man replied while shaking Sam's hand.

Sam smiled, "Hey, I know you're busy so I'll let you get back to work, but I really appreciate your talking with me."

"No problem … thank you. I better go find that harness now," the painter said as he walked away.

✦ ✦ ✦ ✦ ✦

That night at dinner Guillermo told Sam, "I wasn't sure what was going on when you first started talking with the painter, and then I wasn't sure the man was going to like a *gringo* telling him to work safely … you know our culture is different down here."

Sam smiled at that last comment. "I'm not so sure I agree, but let me explain what was happening this afternoon. One of the most important aspects of a strong safety culture is feedback," he began. "I saw the guy on an eight-foot ladder with a tie-off that could easily break if he fell."

"Well, a lot of people aren't too happy when someone tells them they're not doing something the right way. Many of them have been working for years," Guillermo interrupted.

Sam nodded. "We're all like that – it's hard for us to accept feedback, but specific feedback provided in a positive way is effective across all cultures. Let me take you through a *Safety 24/7* Conversation."

Observe

Accentuate

Explore

Emphasize

Agree

Step 1: *Observe* means training ourselves to be more aware of people's behaviors ... both what they are doing safely and what they're doing that might put them, or someone else, at risk."

Step 2: *Accentuate* the positive to lower a person's natural defensiveness, as well as reinforce those safe behaviors we want him to keep doing.

Step 3: *Explore* allows a person to figure out what he did safely or at-risk, which helps him begin to take *ownership* of his behavior.

Step 4: *Emphasize* the consequences of their actions to help people understand the impact an incident could have on them, as well as other people.

Step 5: *Agree* on future actions confirms people understand they are accountable for their behaviors and responsible for their safety, as well as the safety of those around them.

"You've given me a lot to think about," Guillermo responded, "but let me just say this: when I saw you talking to the painter this afternoon, I wondered if you realized we're different down here. Our culture is different, our lives are different ... but after seeing how you handled that situation and hearing what you've shared with me tonight, I'm learning we're not as different as I thought."

Sam shook his hand. "My hope is everyone will start thinking about safe behaviors in everything they do, both on and off the job," Sam said.

Guillermo was thoughtful for a moment. "If everyone could have *Safety 24/7* Conversations like the one I observed this afternoon, it would help us all get home safely to our families, which is something everyone wants to do."

 # SNAPSHOT

DEFINITIONS:

1. *Safety 24/7* **Conversation:** specific feedback provided in a positive way to encourage someone to demonstrate safe behaviors.

ESSENTIAL CONCEPTS:

✦ Risk is determined by the probability of an incident occurring given the associated hazards.

✦ Risk assessing is consciously and purposefully deciding how we are not going to get hurt.

    1. What am I doing?

    2. What might be hazardous about what I'm doing?

    3. What can I do to reduce those hazards?

✦ Since everything we do carries an element of risk and our survival depends on how well we manage those risks, even seemingly simple tasks – such as driving or how we sit at our desks – should be risk assessed.

✦ The more hazards we can eliminate, the lower the risk.

✦ The five steps of a Safety 24/7 Conversation are:

Step 1: *Observe* – being more aware of people's behaviors – what they are doing safely and what they're doing that puts them, or someone else, at risk.

Step 2: *Accentuate the positive* – to lower people's natural defensiveness and to reinforce those safe behaviors we want them to keep doing.

Step 3: *Explore* – allows people to figure out what they did safely or at-risk, which helps them begin to take *ownership* of their behavior.

Step 4: *Emphasize* the consequences of their actions – helps people understand the impact an incident could have on them, as well as other people.

Step 5: *Agree on future actions* – confirms people understand they are accountable for their behaviors, and responsible for their safety, as well as the safety of those around them.

Management's Buy-in — The Foundation of Every Safety Culture

Kurt was eager to meet with his boss after the trip to Venezuela and made sure Sam could attend the meeting as well. "I'm new at this," he told Sam, "so I want you to make sure I dot all the i's and cross all the t's."

"I've been looking forward to this meeting," Ron Kaiser began, after greeting the two men in his office. "As I've mentioned before Bradshaw, these recurring incidents are costing us time as well as money … so, tell me about South America."

Kurt and Sam briefed the manager on what they had seen, their meetings with the employees and their conversations with some of the managers. As they concluded, Kurt changed the focus. "Let's talk about what we need to do to build a strong safety culture for the organization."

"Okay … shoot," Ron said.

"Well, over the last few months I've been seeing things differently. What I've learned is policies, procedures, and good equipment alone don't make a strong safety culture."

"I agree," Kaiser said. "We haven't gotten the results we wanted and our claims continue to rise."

Kurt nodded. "I've shared some of our numbers with Sam, and I think we've come up with a strategy that may turn those statistics around. If we want change to occur, we need to focus on people's behavior and consciously think about what we reward and tolerate."

Kaiser looked interested. "Let's get started on this safety culture thing," he said, resting his elbows on the desk. "People have to realize safety is a priority around here."

Sam leaned forward. "If I can jump in … safety as a priority is what you want to avoid when you're building a strong safety culture."

Kaiser looked confused. "What? Setting safety as a priority lets people know it's important to us."

"Have you told people in the past that safety is a priority?" came Sam's quick reply.

The manager settled back in his chair, folding his arms across his chest. "Well, yes …," he admitted.

Kurt was next to speak. "In a strong safety culture, safety is elevated to be a core value in the organization. It's no longer just a priority."

Kaiser raised an eyebrow. "So, what's the difference? Aren't you just talking semantics?"

"***Priorities change. Core values remain constant***," Sam pointed out. "If safety is only one of many priorities, it typically takes a back seat to cost cutting, operational performance, or client demands. For example, when money becomes tight or operations fall behind schedule the focus becomes getting the job done, no matter what risks or shortcuts you have to take. It's sort of a 'Mission First' mentality. When safety is a core value, the only time it becomes *the priority* is when it comes into conflict with anything else, meaning the organization commits to putting human life above all other demands."

Sam's comments had obviously caught the manager off guard. "Well, uh … now that you bring it up, we do have a lot of competing demands," Kaiser said thoughtfully.

"What we want to do is instill safety as a value … a personal and moral one … within everyone who works for the company," Kurt explained. "Safety isn't a policy people read, remember for a few days, and then forget. Safety is something people practice at work and in their personal lives. Essentially, ***every person is responsible for his or her own safety, as well as the people around them.***"

"Last year we spent over $85 million in claims due to injuries and deaths on the job," Kaiser said, leaning forward and picking up a pencil. "If your so-called behavior-based safety culture can reduce that cost, I'll be the first in line to raise safety from a priority to a core value."

Sam could sense they now had Kaiser's attention. "If the company truly embraces a behavior-based safety approach across the board, you will see a significant decline in claims related to incidents and injury. Using last year's figure, suppose there was only a ten percent reduction in claims. That's $8.5 million directly to the company's bottom line. How much new business is required to generate that kind of profit? More importantly, you'll establish a culture that saves human lives and you can't put a price tag on that."

"Plus, think of the experience we lose when a good employee is out, even for a short period of time. Also, if we create the type of culture that motivates people to work safely, I also think we'll lower turnover," Kurt added.

"How could a safety culture make a difference in employee turnover?" Kaiser wanted to know.

"First of all, in a strong safety culture, the organization rewards safe behavior," Sam responded. "When a manager *catches an employee doing something right*, they praise that person, making the employee feel valued. Also, when an organization elevates safety to a core value, promotions go to those people who practice safe behaviors. That demonstrates to all employees that getting the job done safely is more important than getting the job done fast. Having safety as a core value reduces turnover because people want to work for a company that genuinely cares about their safety."

"Makes sense," Kaiser commented, "But are we going to see management praising employees for safe behaviors all the time?"

Sam chuckled. "That would be a giant step in keeping employees safe."

Now, it was Kurt's turn to speak. "In the type of culture we need to build, everyone would have the right and responsibility to stop a job if they even suspected there's something unsafe about it."

"Whoa … just a minute," Kaiser interrupted. "You mean anybody can stop a job … at any time just because they don't think it's safe to continue?"

"Absolutely right," Sam said. "Remember what we said ... they have not only the right to stop a job if they suspect something is unsafe, they have the *responsibility* to stop the job."

Kaiser shook his head. "I'm not sure I can buy that."

"In a strong safety culture, it's almost a condition of employment that you stop a job if you feel there is anything that puts you or someone else at risk. That could mean stopping the job because you don't understand what's going on. When you think about it, that's the root cause of many incidents," Sam continued. "Let's go back. If safety is a core value – and that means it connects here," he said tapping his chest, "then safety comes first ... period."

"Okay. Okay. I'll have to think about that, but let's talk more about stopping a job ... I need to understand this better."

"What we're saying is that anyone can call a time-out," Kurt emphasized. "If it turns out there was not a problem, then that's okay, they're not going to get in trouble."

Kaiser looked incredulous. "Say that again."

"If someone stops a job because something looks or feels unsafe, or even if he just doesn't understand what is being done, there's no retaliation. They get no grief from their peers or from management. Instead, they get praised for demonstrating the behavior we want ... even if it turned out to be a false alarm."

"Man, that's a huge change." Kaiser's voice was thoughtful as he spoke. "You do realize this is gonna slow down operations – if I have to trust any worker who wants to stop a job?"

Now it was Sam's turn. "Consider how much it slows down operations when there is an incident or someone gets hurt and you lose their experience."

It was obvious Ron was giving serious thought to what Kurt was proposing. "So, let's say we do this. We elevate safety from a priority to a core value. Then, we give people the right and responsibility to stop a job if they see … or even feel … something is unsafe. Nobody jumps on their case if they're wrong. Just the opposite. They are praised for stopping the job and trying to keep the team safe. Right so far?"

"You're right on," Kurt assured.

"Okay, good. So let me go on. What happens if … check that … *when* a manager doesn't support this new culture?"

Kurt's voice remained confident. "If they don't support the safety of their personnel, or they don't trust someone to stop a job to keep everyone safe, are they the people you want on the team? Any manager – even the most senior or productive person on the management staff – who tolerates at-risk behaviors will eventually either change his perspective or the new culture will squeeze him out. You see, if the executive team is pushing down from the top, and the people at the sharp end of the stick are pushing up from below, there is nowhere for those managers to hide."

Kaiser's expression was one of shock. "We've got some managers who do a real good job for us … and have done so for a lot of years," he countered.

"He or she isn't doing a good job if they're turning a blind eye to at-risk behavior and rewarding shortcuts," Kurt said. "All that manager cares about is how fast the job gets done. But Ron, you know as well as I, those short cuts are exactly the behaviors that get people hurt."

The manager was reluctant to agree. "That's a pretty tall order you're throwing out."

Sam maintained a poker face. "Or, the company can keep doing what it has already been doing with regard to safety. But to expect different results is my definition of insanity."

Again, Kaiser was silent as he looked at the photographs of his wife and two sons featured prominently on his wall. "I see what you mean," he murmured.

Kurt gave Ron a few moments to think before continuing. "While we were in Venezuela, Sam had a conversation with one of the men at a job site. The man was working on an eight-foot ladder and hadn't tied off with a proper safety harness. After Sam spoke with him, that guy was not only willing to change the way he was doing his job, I think he would even speak up if he saw someone else doing something at risk. To get the results we want, employees need to understand *feedback is the breakfast of champions*, even if it means having a difficult conversation," Kurt concluded.

"That kind of training will take time," came Kaiser's response, "and what will it cost?"

"You're right," Kurt said, "and the sooner we start, the sooner we'll be able to realize those savings we discussed earlier, which will more than cover the cost of training. But we're going to need management's commitment and unconditional support every step of the way."

"What specifically would management need to commit to?" Ron questioned.

"Glad you asked," Kurt grinned and showed Ron a list he had prepared before the meeting. "Here's what we need to instill a strong SAFETY culture in the day-to-day operations of the company."

Support safety as a core value by committing to put human life ahead of all other demands.

Accountability gives all employees the right/responsibility to call a time out and rewards them for doing it, even if it's a false alarm.

Follow-up by demonstrating and communicating a personal commitment to safety in all of your actions.

Elevate people who support the new culture and eliminate those who tolerate at-risk behavior, even top producers.

Train our people to observe at-risk behaviors and have *Safety 24/7* Conversations.

You are the key to an incident-free environment.

It was obvious. Ron Kaiser was not only impressed with Kurt's approach, but he was also beginning to see the possibilities. "You know, Bradshaw, I've been around for a long time, and I've seen a lot of safety programs come and go. Every one of them failed to keep our people safe. But, this program of yours …."

"It's not a program," Sam interjected. "A program has a distinct starting and stopping point. *If you want to maintain a strong safety culture, you have to continually reinforce it and make sure it stays relevant as the company evolves.*"

Kaiser cleared his throat. "It's apparent I'm going to have to change my thinking about this whole safety issue," he said, in a strange way apologizing for his lack of insight. "But, I think it's something we should try here."

Kurt was pleased with his boss's response. "I agree."

Kaiser smiled for the first time since the meeting had begun. "I'm willing to help our people move safety from their heads to their hearts," he said. "The future of our company belongs to those who have the courage to call a time-out, the ability to speak up, and the willingness to listen."

Sam looked at Kurt. "I think Ron is ready to make this journey with us."

 # SNAPSHOT

ESSENTIAL CONCEPTS:

✦ In a strong safety culture, safety is elevated to be a core value in the organization. It's not just a priority. (Priorities change. Core values remain constant.)

✦ Having safety as a core value reduces turnover because people want to work for an organization that genuinely cares about their safety.

✦ Policies, procedures and equipment alone don't make a strong safety culture.

✦ All people are responsible for their own safety, as well as the safety of the people around them.

✦ To change a culture, focus on peoples' behaviors, consciously rewarding safe behaviors and not tolerating at-risk behaviors.

✦ In a culture of safety, each of us has the right and responsibility to stop a job if we even suspect there's something unsafe about it.

✦ Everyone – even the most senior manager or productive person – who tolerates at-risk behaviors will eventually either change their perspective or the new safety culture will squeeze them out.

✦ In a strong safety culture, we understand feedback is the breakfast of champions, even if it means having a difficult conversation.

✦ To demonstrate management buy-in to a culture of *Safety 24/7*:

S*upport* **safety as a core value by committing to put human life ahead of all other demands.**

A*ccountability* **gives every employee the right and responsibility to call a time out and rewards them for doing it, even if it's a false alarm.**

F*ollow-up* **by demonstrating and communicating a personal commitment to safety in all of your actions.**

E*levate* **people who support the new culture and eliminate those who tolerate at-risk behavior, even top producers.**

T*rain* **your people to observe at-risk behaviors and have** *Safety 24/7* **Conversations.**

Y*ou* **are the key to an incident-free environment.**

GIVING AND RECEIVING FEEDBACK

After his meeting with Ron Kaiser, Kurt was ready to begin implementing the new *Safety 24/7* culture. Realistically, he didn't expect changes to occur overnight. He knew it would be built one step at a time.

"I'm going to unveil the concept at the supervisors' meeting today," he told Jessica over breakfast that morning. "I am going to stress to them the importance of giving and receiving feedback."

"Communicating has always been one of your strengths," his wife responded. "You recognize ***listening is more than just waiting for your turn to talk.***"

Kurt hugged his wife before heading for the door. "Wish me luck today."

✦ ✦ ✦ ✦ ✦

Kurt felt good about his agenda for the meeting. After a few icebreakers, Kurt was ready to introduce some of the key components for the new safety culture.

"It's called a *Safety 24/7* culture for several reasons," he began. "First, the word *culture* describes the behaviors and beliefs of a group of people that are transmitted from one generation to another. In an organization, that means what experienced personnel pass along to new hires. With our international operations, the term culture also refers to the many countries represented by our workforce.

"Secondly, when we are truly committed to safety, we demonstrate safe behaviors in all aspects of our lives 24/7. Safety isn't something we 'turn on and off' at work."

Several heads nodded in agreement.

"An important concept I've learned," Kurt continued, "is in a culture of safety, feedback is the tool our teammates have for reminding us when our behaviors impact safety.

"Think about a person who gave you good feedback – feedback that motivated you or helped you see a situation for what it really was. What made that particular feedback so effective?"

"I accepted feedback better if someone described *what* they saw me doing," said one supervisor after a few moments. "But, I didn't like it when they tried to tell me *why* I had done it that way: for example, if they said I was just being lazy or had a bad attitude."

"Exactly. Effective feedback focuses on actions that can be observed, not attitudes, because we can never really know what someone else is thinking," said Kurt. "Good point! Someone else?"

A supervisor sitting at the far table smiled. "Many years ago, our minister sat me down and talked to me straight. He was very direct and specific on what I needed to do … I could tell he truly believed what he told me. I'll never forget him or what he did for my life."

"The feedback I remember was from my drill sergeant," came another reply. "He could have made me feel like the lowest person in the world. Don't get me wrong. He was in my face, but he also praised some of my good qualities and said he believed in me."

"These are all great examples," Kurt applauded, "and show you already know effective feedback is based upon a sincere desire to help someone improve. Plus, it focuses on what the person is doing right, as well as what he could do better."

"Okay, I've got a question for you," a participant asked. "From a safety standpoint, I've often seen people do something risky because they think, 'I can get away with it.' What's the best way to handle that?"

Kurt immediately thought of his first meeting with Sam. "We have a name for those behaviors," he said. "We call them *old school* or *bulletproof* because the person doesn't think they, or anyone else, will ever get hurt by their actions. But let me ask you something. What did you do when you saw someone take that risk?"

"Well, they got the job done, so I didn't say anything," came the reply.

"You're not alone in handling the situation like that. We've all done it – but what we're doing when we ignore these behaviors is actually rewarding and encouraging the employee to take risks in the future. People are essentially practicing for their next incident. In a *Safety 24/7* culture, we need to provide that person feedback to help him change his behaviors."

"But isn't the other person's behavior out of our control?" asked a supervisor.

Kurt was thoughtful before he responded. "Yes, ultimately a person is responsible for his own behavior, but what I have to ask myself is, 'Have I provided him feedback in a manner that motivates him to change his behavior?' If not, then I am accountable if an incident does take place, which is why providing effective feedback is so important."

✦　✦　✦　✦　✦

Kurt started the second part of the meeting by writing, '**Trust is essential in a culture of safety**,' on the board and asking the question, "How many of you trust everyone you work with?"

He waited for a moment or two. Not a single hand went up. "Okay, let me ask another question. What does it mean when you trust somebody?"

"To me, it means having confidence in that person," came a reply.

"It means they've delivered on what they promised in the past," said another supervisor.

"Your answers are right on," Kurt responded. "Trusting someone means believing in them and what they say. It's knowing that person is sincere."

"My wife trusts everyone. For me, people have to earn my trust," a supervisor spoke up.

"Good point," Kurt said. "Whether or not you immediately trust someone, a key step in building a culture of safety is to *commit* to building trust. Do you risk being deceived or disappointed? Absolutely.

There are risks involved in making this commitment because, as we all know, sometimes trusting a person backfires – and the results leave us feeling ripped off and vulnerable. However, if we don't start trusting each other, the long-term consequences are much worse.

"So, in order to build our *Safety 24/7* culture, I'm asking you to make a sincere commitment to trust the person providing you the feedback … and here's what you've got to do to fulfill the commitment," he said, pointing to a PowerPoint slide:

Building trust requires your willingness to:

- ✦ Recognize the person may see things differently than you do (and you can still both be right)
- ✦ Be receptive (you may not know everything)
- ✦ Listen (not say, 'I know.')
- ✦ Ask questions
- ✦ Focus on areas of agreement
- ✦ Act on the feedback
- ✦ Follow-up

Kurt gave the group a chance to absorb the information.

"In a Safety 24/7 culture, trust throughout the organization is critical," Kurt said. "Let me ask you, has anyone been to a marine park that has a show using killer whales?"

A few of the supervisors slowly raised their hands.

"Bet you're wondering what killer whales have to do with feedback and trust," Kurt chuckled. "Considered one of the most feared predators in the ocean, killer whales weigh several tons and can kill just about anything. *Whale Done!*, a book by Ken Blanchard, tells about the trainers at Sea World who not only get into the tanks with these giants but also ride on their backs. The book details how the trainers motivate whales to perform, and the first thing Blanchard talks about is trust — not just trainers trusting the whales but the whales trusting the trainers.

"How many of you would want to get in the tank with a killer whale that had just been punished for not doing what it was supposed to do? I know I don't respond very positively when people yell at me."

"Me either," said one of the participants.

"When a whale does what the trainer wants, it is praised and rewarded. If the whale gets off track, the trainer sets up another objective — maybe a repeat of the last one — so the whale can be successful and get rewarded," Kurt explained. "So how can we apply this to our own people?"

"I guess what you are saying is, if someone does something at-risk, it's not a good idea to yell at them or to ignore it. Instead we need to redirect them in order for them to be successful," one of the supervisors volunteered.

"Exactly," Kurt affirmed. "And why is it not a good idea to simply tell someone 'Don't do that again?'"

"Because there is a reason why the person performed the way they did, and unless we understand what motivated them to do it that way, we shouldn't expect their behavior to change just because we want them to," a supervisor explained.

"You guys are quickly understanding the logic behind the *Safety 24/7* culture," Kurt praised.

✦ ✦ ✦ ✦ ✦

By the end of the week, Kurt was sure he was on the right track. His meetings had gone as well as he had expected, and Kurt believed he had earned buy-in from several of his team.

"They've been willing to listen, once we discussed the impact our safety statistics have on their operations," Kurt said as he helped Jessica clear the dinner table. "I will admit, my expectations have been pretty high, but so far, I haven't been disappointed."

"You've always operated with a positive attitude and high expectations," his wife replied supportively. "Remember how we were the summer before Shannon went into middle school? We were sure she'd shut us out of her life, and we wouldn't know any of her friends."

"Oh man, I lost lots of sleep over that," Kurt remembered.

"But, we trusted Shannon and believed she would make good decisions," Jessica continued. "By having those expectations and rewarding the good choices she made, Shannon never let us down ...."

"Well, with volleyball and softball, she doesn't really have time ... and after all, she's her daddy's girl," Kurt chuckled.

"You're right," Jessica agreed, "but trust along with high expectations can make a big difference."

When Kurt arrived at Sam's the following day, he was surprised to see two other men.

"Kurt, I wanted you to get to know these guys," Sam began the introductions. "I've told them about your task and what you've been doing over the last 60 days. Both are experienced hands and since you mentioned you wanted to talk about feedback, I thought it would be good to hear their comments. Meet Cy Thomas and Mark Wilson."

The two men stood and shook Kurt's hand. Both appeared to be in their early-60s; with weathered faces and calloused hands, Kurt knew they'd spent a lot of time working outdoors.

After briefing the trio on the high points of his meetings during the past week, he added, "Some of the supervisors voiced concerns about their people wanting to provide feedback to their colleagues."

Mark Wilson was the first to speak up. "I can see their point. It's tough to offer feedback to people, especially co-workers who have been on the job longer than you. However, if you can help people understand you genuinely care about their safety, they're usually more willing to listen.

"But, giving feedback is only half of it. People also need to learn how to receive feedback effectively. I've written down a few things that have helped me over the years:

- ✦ *I sincerely listen to the feedback people offer.*
- ✦ *I separate what the person says from what I think about the person.*
- ✦ *I don't overreact to feedback.*
- ✦ *If I'm not clear about what they're saying, I ask open, non-defensive questions.*
- ✦ *Even when I disagree with parts of the feedback, I focus on areas where I can improve.*

Kurt reviewed the list. "Those are all great points," he said.

Cy Thomas was a slight, wiry man who spoke with a definite drawl. "The important point for people to remember is the consequences of not offering feedback can be devastating … it can literally be the difference between life and death."

"So, is it harder to give feedback than to receive it?" Kurt wanted to know.

"Depends on your attitude. Think about the last time your wife gave you feedback. Let's say you were working under the hood of the car and your wife came out and made a suggestion – like using gloves or taking the keys out of the ignition," Cy said, looking away to avoid Sam's steely gaze. "How would you respond?"

Kurt thought back for a moment, then acknowledged sheepishly, "If I was in a hurry, I might say something like, 'If you don't like the way I'm doing this job, perhaps you'd like to do it yourself.'"

The others laughed. "Me too," Cy admitted. "But, if I believed my wife genuinely cared about me, I would realize she was saying it for my own well-being, and I would be more likely to follow her advice."

"It also depends on the other person's history of providing you feedback," Mark pointed out. "If your wife was what I call a 'seagull,' only swooping in to dump negative feedback on you all the time, maybe you wouldn't listen as closely or do what she suggested."

"The same is true with employees," Cy said. "If you're on their backs constantly, how open do you think they'll be to your feedback, even if it could impact their safety? It's important for people to receive feedback on an ongoing basis, but the majority of it should be about things they're doing right. Besides, telling someone 'I told you so' can never be considered positive feedback.

"Remember, it's not just supervisors who give feedback. Everyone does. In fact, co-workers are more likely to be around when opportunities for feedback come up, since there are more of them. But, it takes courage for team members to speak up. That's why building trust is the first step."

"Let me add one thing here," Sam paused and took a deep breath. "I've been in situations where honest feedback was given, someone chose to ignore it, and a serious injury happened. So there's always the question, once honest feedback is given, will the person choose to change his behavior?"

The room was silent. Finally Kurt spoke, "Like you said in our first meeting, we need to help our people take that information from here (pointing to his head) and move it to here (tapping his heart)."

Sam smiled. He knew Kurt had learned this lesson well.

SNAPSHOT

ESSENTIAL CONCEPTS:

+ Feedback is the tool our teammates have for reminding us when our at-risk behaviors impact safety.

+ Effective feedback is based upon a sincere desire to help someone improve and focuses on:

    1. Actions that can be observed, not attitudes, because we can never really know what someone else is thinking.

    2. What the person is doing right, as well as what he could do better.

+ Ultimately everyone is responsible for their own behavior, but we still need to ask ourselves, "Have I provided them feedback in a manner that motivates them to change their behaviors?" If not, then we are accountable if an incident takes place.

+ In building a *Safety 24/7* culture, if you can help people understand you genuinely care about their safety, they're usually more willing to listen to feedback.

+ Trust is essential in a culture of safety.

+ We shouldn't expect coworkers' behaviors to change just because we want them to.

+ We can learn how to receive feedback effectively by:

    • *Sincerely listening to the feedback people offer (and remember, listening is more than just waiting for your turn to talk).*

    • *Separating what the person says from what I think about the person.*

    • *Never overreacting to feedback.*

- *Asking open, non-defensive questions if I'm not clear about what they're saying.*

- *Focusing on areas where I can improve, even when I disagree with feedback.*

✦ The consequences of not offering feedback can be devastating … it can literally be the difference between life and death.

✦ It's important for people to receive feedback on an ongoing basis, but the majority of it should be about things they're doing right.

✦ There's always the question, once honest feedback is given: Will the person choose to change his behavior?

MAINTAINING MOMENTUM

Back in his office after a whirlwind schedule of training sessions with supervisors and employees around the country and overseas, Kurt was cautiously optimistic about his goal of creating a culture of safety for his organization.

In the past, safety had been a priority until something more urgent came up. It was becoming clear. Building a strong safety culture required everyone in the company, from the president to the newest hire, to understand the importance of making safety a core value, not just a priority that changes as needs arise or deadlines appear.

At the various job sites he visited, Kurt assured his people, "It's not a new initiative. The whole emphasis is on taking the good things we are doing and ingraining them in our culture so everyone 'walks the talk' 24/7."

Kurt remembered a comment from a supervisor who had been around for many years: "You can talk to me about safety all you want. Just don't tell me how to do my job."

"Talk about *old school* mentality," Kurt mumbled as he switched on his computer. "That attitude will keep people from speaking about their concerns. **You can't be open to safety feedback if you aren't open to all feedback."**

Thinking more about that supervisor, Kurt began typing an email to all the supervisors he had been training. He titled it: *"Maintaining an Incident-Free Culture."*

"During the last 90 days, many of you have helped me learn what it takes to build a strong safety culture. You've helped me realize in order to build the culture we want, **we must consciously think about what our actions demonstrate, reward and tolerate.**

"It's not easy, and it's not something that happens on autopilot because our normal actions usually reinforce old habits, which only helps to maintain the current culture. Here's an example:

"Consider an employee who takes a shortcut to get a job done. If the supervisor rewards the end result without realizing the shortcut involved at-risk behaviors, the supervisor has, unconsciously, rewarded those behaviors.

"As supervisors, we also need to consider what we *tolerate* in our employees. Is it possible we overlook certain at-risk behaviors because it is easier to ignore them than to deal with them? Do we make excuses for our employees – and for ourselves – by saying, 'Oh, that's just how Joe does things,' or, 'I don't have time to deal with it now. Besides HR never does anything, even if I do write up an employee who's put himself – or others – at risk.'"

"I want you to think about these points and then answer the following questions:

+ What do your actions say to other people? Do you continually demonstrate safety is a core value to you?

+ What behaviors do you reward in other people? Do you only look at the outcome of a job and not the risks taken to get it done quickly?

+ What behaviors do you tolerate? Do you find excuses not to speak up?

Kurt ended his email by reinforcing the need for praise in a changing culture. He also included suggestions about how to praise performance from Ken Blanchard's book, *Whale Done* – called "*The Whale Done Response.*"

+ Praise people immediately

+ Be specific about what they did right ... or almost right

+ Share your positive feelings

+ Encourage them to keep up the good work

The following morning, responses began to come in.

One supervisor wrote:

"There are several employees who think this whole thing is just another feeble attempt to try and polish the organization's safety record ... and they don't think it's worth making any real changes because they believe it will go away, just like past safety initiatives have. At any rate, they're giving me fits and may never get to the place where they take it seriously. I feel like a failure ... any ideas?"

"Yeah," said Kurt to himself. "I've seen that resistance at every job site I've visited and it isn't pretty."

He emailed the supervisor this response:

> "Cultural change is never easy. You are asking people to change the way they've been doing things for years. The old culture will fight it at every opportunity.
>
> "When trying to change a culture, it is natural for us to spend time with the minority of people who resist change. That's NOT where we want to spend our energy. Those 'resistors' are not going to change quickly.
>
> "Instead, team-up with the 'supporters' of change. This helps in two ways: first, the supporters are the champions you need to keep on board. They have a tough role and can easily get discouraged because they're the ones on the front lines, taking all the flak. They need your frequent – and visible – support to know they've made the right decision ... plus hearing you reward the supporters will help others get off the fence and align with the supporters. Getting the 'fence-sitters' on board ... combined with your tough love ... will provide the momentum necessary to change the resistors.

But, out of all the emails Kurt received, the one that moved him most was the one that had been written from the heart:

> I could have saved a life that day,
> But I chose to look the other way.
> It wasn't that I didn't care,
> I had the time, and I was there.
> But I didn't want to seem a fool,
> Or argue over a safety rule.

I knew he'd done the job before,
If I called it wrong, he might get sore.
The chances didn't seem that bad,
I have done the same, he knew I had.
So I shook my head and walked on by,
He knew the risk as well as I.
He took the chance, I closed an eye,
And with that act, I let him die.

I could have saved a life that day,
But I chose to look the other way.
Now every time I see his wife,
I'll know I should have saved his life.
That guilt is something I must bear,
But it isn't something you need share.
If you see a risk that others take,
That puts their health or life at stake.
The question asked, or thing you say,
Could help him live another day.

If you see a risk and walk away,
Then hope you never have to say,
I could have saved a life that day,
But I chose to look the other way.

✦ ✦ ✦ ✦ ✦

Later that afternoon Kurt picked-up his daughter, Shannon, from softball practice. "How was practice today?" Kurt asked as Shannon fastened her seat belt.

"It was pretty cool," was the teenager's enthusiastic reply.

Since her team had been struggling recently and the coach was initiating several changes on the field, Shannon's response was a welcomed surprise for Kurt. "Yeah, how come?" he asked.

"Well, coach started the practice by reminding everyone of our goal to win the division championship. Then he talked about why he had made some specific changes and said no matter how strong a team is, he knows each of us will react differently when changes are introduced," Shannon said taking a piece of paper out of her workout bag.

"Coach said change typically causes people to have concerns like ..."

- *How will the change impact me personally?*

- *What's in it for me?*

- *Will I win or lose?*

- *Will I look good?*

- *Will I have enough time to learn my new position?*

- *Can I do it?*

- *What if I don't like it?*

"Those are all good questions," Kurt agreed. "So then what happened?"

"Well, first it just made me feel better, knowing my worries were normal. I was beginning to wonder if I would even be playing. Then coach sat down with each of us and talked about our individual strengths. He helped me see how my past experience and success could help me be successful again ... and he gave me some new stuff to think about, too. Now I understand why he made the changes and, more importantly, how those changes will affect me ... so now I can concentrate on my new role on the team," Shannon concluded with a smile.

"Your coach sounds like he really knows how to lead people. Mind if I borrow that list?" Kurt asked, thinking of the team he was trying to build.

✦ ✦ ✦ ✦ ✦

The following day, Kurt went by Sam's house to bring him up-to-date on his activities. "Sam, I think everything is going pretty much the way I envisioned, thanks in no small part to you," he said. "I also think a majority of our employees are excited about building a culture of *Safety 24/7*. Now, one of my main concerns is how to keep the momentum going."

Sam smiled. "You've just been through the birth and delivery of your safety culture. Now you want to make sure it has the opportunity to grow?"

"That's a pretty good analogy," Kurt agreed.

"I want to share something I read several years ago about maintaining momentum." Sam pulled a folded piece of paper from his pocket, smoothing it so Kurt could read the words:

When changing a culture:

    (1) Don't expect to be popular

    (2) Recruit help

    (3) Educate and train

    (4) Champion change at every opportunity

    (5) Measure and reward results

    (6) Have the courage to stick with it

"Hmmm … some good ideas here," Kurt said, "but I'd like to talk about the details."

"Glad to," Sam said. "Let's begin with:

1. Don't expect to be popular.

"In any culture change, that small group of people who want to hold on to their old behaviors will use every opportunity to discredit both you and the new culture you are creating. Remember, you're doing what you sincerely believe is necessary to help the organization and the people. You're not out to win a popularity contest."

Kurt shook his head. "That's pretty straight forward," he agreed.

"Moving on to:

2. Recruit help.

"Achieving a strong safety culture requires a team effort. It doesn't merely have to be supervisors talking with employees. When team members begin talking to other team members about safety behaviors, it doesn't matter if the supervisor is around or not ... the safety culture will become self-supporting because there's almost always another team member around to have those important *Safety 24/7* Conversations."

"Got it. The supervisor doesn't have to carry the entire load himself," Kurt said. "Just like I mentioned earlier ... it makes so much sense when you lay out the big picture. Keep going. This is good stuff!"

3. Educate and train.

"This means educating employees about why change is necessary and then working with them to develop new behaviors. For example, the training you're doing on how to give and receive feedback," Sam said. "Until everyone knows how to do it, they'll stay in their comfort zones and stick with the old way of doing things."

4. Champion the change at every opportunity.

"Employees watch their leaders to see how committed they are to the culture change," Sam continued. "Communicate the successes. Tell stories about how people are meeting the standard. Talk about the difference in their incident numbers and compare how they are doing now with how they were doing before the change took place."

"Sort of like a cheerleader?" Kurt asked.

"Cheerleading is certainly part of it, but so is motivating, teaching, and – since you're down in the trenches too – leading by example. I like to think of it more as a 'player-coach' scenario. Make safety a part of every meeting, every conversation. Any time it's possible, talk about improving your culture of safety."

"I'm ready," Kurt said.

5. Measure and reward results.

"Decide what you will measure and then communicate those measurements and make the results known immediately. Short-term reporting helps employees see you're serious about tracking the changes that are needed," Sam said. "Many organizations fail to follow up. Once employees realize you're totally committed to building a strong safety culture, more fence-sitters will be convinced to join the early adaptors."

"Okay," said Kurt.

6. Have the courage to stick with it.

"How do you think a supervisor feels after a day of beating his or her head against the wall, trying to convert resistors? Now think about how they'd feel after a day of working with the early adaptors who support the new safety culture. Remind your supervisors – it feels a lot better rewarding those early adaptors. It's less stressful catching people doing something right," Sam pointed out, "and you know from your own experience that changing a culture creates a whole new brand of stress."

Kurt nodded his ready agreement.

"There will always be plenty of push-back from the resistors, always a lot of people questioning what you're doing," Sam empathized. "Sometimes, you'll even question yourself about whether or not you're doing the right thing."

"Been there a lot lately," Kurt admitted.

"I'm sure you've also had times when you felt it would just be a lot easier to chuck the whole process and go back to the way things were … but don't do it. It takes courage to stand your ground and not back down for an easier journey, and you need to regularly remind yourself why you're doing this.

"Whenever the going gets tough, you'll have to imagine how good it's going to feel – for you, all your people, and their families – when an incident-free work environment is achieved."

 # SNAPSHOT

ESSENTIAL CONCEPTS:

+ You can't be open to safety feedback if you aren't open to all feedback.

+ To build a strong safety culture, we must consciously think about what our actions demonstrate, reward and tolerate.

+ To build a culture of Safety 24/7, you are asking people to change habits. Expect the old culture to fight the culture change at every opportunity.

+ Team up with the 'supporters' of change to help the 'fence-sitters' get on board, which will provide the momentum necessary to change the 'resistors'.

+ When making changes, people will always have individual concerns:

    (1) How will the change impact me personally?

    (2) What's in it for me?

    (3) Will I win or lose?

    (4) Will I look good?

    (5) Will I have enough time to learn my new position?

    (6) Can I do it?

    (7) What if I don't like it?

+ When people can honestly say, 'I believe in safety for myself and actively encourage my teammates to believe in safety,' their passion for safety comes from their heart, not just their head.

✦ When changing a culture:

    (1) Don't expect to be popular

    (2) Recruit help

    (3) Educate and train

    (4) Champion change at every opportunity

    (5) Measure and reward results

    (6) Have the courage to stick with it

KEEPING IT REAL

"Two weeks to go," Kurt sighed as he glimpsed at the calendar on his desk. But, he felt good about the progress they had made over the last 3 ½ months.

"We're not starting another new program," he repeated time and again as he traveled around the country and down to South America. "We want to strengthen our existing safety culture by addressing the human element."

In some cases, his approach was predictably met with resistance, but momentum started to grow as more people began taking responsibility for their own safety, as well as those around them and, more importantly, incident rates started coming down.

His ringing phone jolted Kurt out of his thoughts.

"Ron? Sure! I'll be right up," he said.

His boss had sounded rattled. When Kurt walked through the door, he found Ron staring at a map on the wall behind his desk.

"Glad you were in town," Kaiser began, his words coming in rapid-fire succession. "We've just had a serious incident in Saudi Arabia. We need to begin focusing our efforts there. But with the cultural differences, I'm not sure this new process will work in that part of the world. Maybe we need to try something different."

Kurt listened as Kaiser provided the details and thought about his answer before he spoke, making certain his voice remained calm but assertive. "I am confident *Safety 24/7* is powerful enough to transcend geographic and cultural differences because, as hard as it may be to believe, human behavior is the same wherever we go."

"What about different religious beliefs, as well as various customs and traditions?" Kaiser wondered. "They don't think about things the same way we do."

Kurt was ready for Ron's questions. "What we need to focus on are those things people have in common no matter what country they are from. Think of it this way ... if we line up five people from different cultures and I walk by and stomp on each of their feet, tell me in what culture that wouldn't hurt?"

Kaiser seemed to be searching for the right answer. "Well ... uh ... of course it would hurt," he said finally.

"Exactly," Kurt countered. "A person's desire not to get injured is universal. There are other things we all have in common as well, like the bond of family, the need for food, the desire to remain employed and the need for respect.

"Here's an idea ... why don't we call John Sullivan and talk with him about what he's encountered with the multiple cultures he has in South America?"

Kaiser agreed and hit the speed dial on his phone. "Sullivan? Ron Kaiser. I've got Kurt Bradshaw here, and we need your thoughts on our new safety culture. We've got you on speaker phone."

"I'll be happy to give you my take," John said, his Australian accent apparent in every word. "G'day, Kurt."

"Keeping out of trouble down there?" Kurt joked.

"Yea, I'm toeing the line," John retorted, "and Theresa is keeping an eye on me, as well."

Then his tone turned serious. "So, what do you want to know about safety in South America – because I know you've seen some of the positive results."

"I need to know why you think it's working," Kaiser said. "What are you seeing, especially among the different nationalities working together down there?"

After a moment John responded. "What I'm seeing are supervisors taking time to mentor, encourage and congratulate their teams on the job they're doing. I'm seeing people walk a little taller, with more confidence and, best of all, employees are seeing coworkers take care of each other, regardless of where they're from.

"What's happened is people are actually having *Safety 24/7* Conversations and calling timeouts. That, alone, will drive down our incident rate. People like our new safety culture because it matches the desire every worker has to go home from work in the same shape he was in when he arrived."

Kaiser still looked skeptical. "Tell me what else is different?"

"Well, we've developed a group of safety champions – people who are leading the charge ... and in addition to encouraging safety conversations between all levels of staff, we've made sure to reward behaviors we want to see," John continued. "Also, as you know, several weeks ago we fired one of our senior managers ... a guy who consistently turned a blind eye to at-risk behaviors ... which demonstrated to everyone that nothing is more important than safety."

Kurt smiled and gave Kaiser a big thumbs up. Kaiser nodded his agreement.

John continued, "But what's been really great is our people are doing a better job of assessing risks, regardless of what language they speak. Of course, there are still those instances where someone will take a shortcut or do something at-risk, but most of the time, there's someone in the area who stops them before they might hurt themselves or someone else."

"Okay," Kaiser interrupted. "So, do you think what you're doing in South America can work in the Middle East?"

"If it works here," John said, "it can work anywhere. Remember, we have a real diverse workforce. Depending upon the country, we'll have multiple languages on most shifts. But our workers all share the need to stay healthy, get a paycheck and feed their families."

"We may need to adapt *how* we deliver the message, or the methodology of the training," Kurt pointed out, "but what John is saying is people are people, anywhere you go, and ***a person's underlying motivations for working safely are universal.***"

Kaiser looked at his watch. "Well, you guys convinced me this can work across different cultures. I'm astounded at the progress," he said. "I have five minutes to get to a meeting, so we have to end it here. Sullivan, you're doing a great job down there … and we're glad to have you back at the wheel. Bradshaw, let's get started in the Middle East."

✦ ✦ ✦ ✦ ✦

Monday night, Kurt and Jessica invited Sam out to dinner to thank him for all of his help.

"I've heard so much about you, I feel like I already know you," Jessica said to Sam after her husband had completed the introductions.

"And I can't tell you how much I've wanted to meet Kurt's partner – and, if I heard him correctly, his better half," Sam replied. "I had hoped my wife Edith would be able to join us, but she was called away suddenly to take care of her mother. I hope you don't mind Kurt and me talking business for a few minutes; I'm always eager to hear what's been happening with the new safety culture."

"Everything so far has been working well with only a couple of glitches along the way," Kurt began. Through Ron's support of *Safety 24/7*, we've gained momentum and have upper management's buy-in."

Jessica jumped in. "Kurt makes it sound easy, but trust me, it sure didn't start out that way."

"How so?" Sam wanted to know.

"I'm not sure I shared with you how nervous I was about taking this job," Kurt explained.

"I think your exact words were something like, 'What do I know about safety? It's entirely different than operations,'" his wife recalled.

"Jess is saying exactly what I thought," Kurt continued. "I couldn't see how being successful in operations would help me improve safety.

"Over time, though, I began to realize that regardless where I worked in the company, I needed to:

- ✦ *Build a foundation of trust.*
- ✦ *Set clear goals and high expectations.*
- ✦ *Praise progress as people begin changing behaviors.*
- ✦ *Have the courage to keep going.*

"Sounds like you now realize the skills necessary to be a good safety leader are the same skills necessary to be a good leader … period," Sam stated.

"If I had recognized in the beginning, **operations and safety are 'interdependent'**, I could have saved myself a lot of anxiety."

"When you were in operations, if anyone had told you **safety is as much your responsibility as it is for the people who work in the safety department**, would you have believed it?" Sam asked.

"No, probably not," Kurt admitted. "But now I realize safety supervisors are here to *support* us, and it is not their job to keep everyone safe, anymore than it is mine."

"Hey, is anybody hungry besides me?" Jessica interrupted as she picked up her menu. "If we don't order soon, they might ask us to leave."

Once the waiter had taken their orders, Kurt was ready to cut to the chase. "Sam, one of the reasons we wanted to have dinner tonight was to thank you for everything you've done to help me the past four months. I want you to know, my friend, how grateful I am for your encouragement and patience, as well as your willingness to listen and provide guidance every step of the way."

Then it was Jessica's turn. "Kurt said you were one of those people who had everything, but we want to give you something to let you know how much we appreciate all you've done." She handed him a gift-wrapped box.

Inside was a set of bookends Kurt had asked a local artist to make out of safety gloves.

Sam studied the attractive bronze pieces. "Thank you," he said in a strangely soft voice. "You don't know the amazing irony of your gift."

Kurt set down his iced tea glass and leaned forward to hear as Sam told his story.

"Years ago, when my son Jason was just a kid, he and I were getting ready to go to a baseball game when our car's engine made some strange noises. I popped the hood and took a look to see if I could spot the problem.

"About the same time, my wife came out on the back porch to see why we hadn't left. Edith asked if she could help and, as you might expect, I told her 'No, I can handle it myself.'"

"A few minutes later, she brought me a pair of heavy gloves I sometimes used when tinkering with the car. Of course, I knew everything at that point in my life and had been around cars for years. 'Edith,' I said impatiently. 'I don't need the dang gloves.'

"As she left, Edith said, 'Jason, get out of the car while Dad is working on it.' But I fired back, 'He's fine. I won't be a minute.'"

Sam paused in his story long enough to take another sip of tea. "Remember when we talked about how sometimes honest feedback is given and people choose to ignore it?"

Kurt nodded.

"Well, my bullet-proof attitude got in the way of Edith's honest feedback. Two minutes after she gave up and went inside, I learned safety happens by choice ... and so do incidents."

"The noise I heard in the car was my fan belt slipping. I was fiddling with it when Jason got tired of sitting in the hot car and turned on the ignition to start the air conditioner."

Jessica covered her mouth, almost visualizing the outcome.

"I was in too big of a hurry that day ... anyhow, when the engine started, my hand was caught in the fan blade, well ... things got pretty bad.

"Edith heard my screams and came running. When she saw what had happened, she had the presence of mind to turn off the engine. In the meantime, my hand was a mangled mess.

"I was fortunate – I only lost part of one finger, but for a while it was anybody's guess about how much of my hand could be saved. Luckily, I had a good doctor and after a couple of surgeries, followed by several months of therapy, my hand worked pretty well," Sam demonstrated flexing his fingers.

"Oh, how horrible," Jessica said.

"But the physical injuries weren't as bad as the other problems I caused that day."

"You caused?" Kurt asked.

"Jason was about 10 at the time and blamed himself for his dad being hurt," Sam continued. "He developed some pretty serious emotional problems, and several times he ran away.

"Finally, we found a psychologist to work with him so he wouldn't continue to carry so much guilt," the older man explained, "but from the day of the incident, and for many years afterwards, Jason wasn't comfortable with me. It drove a real wedge into our relationship."

"What about Edith?" Jessica wondered.

"I really wish she could have been here tonight," Sam said. "She's an amazing lady, but after the incident, she also felt guilty … and I wasn't very supportive, I'm afraid.

"I went through a period of deep depression during my recovery, which put a real strain on our marriage. I wasn't able to work and money became a big issue," Sam continued. "Finally, it just got to be too much and we separated for several months. I suppose it took that long for me to come to my senses and realize I wasn't as tough as I had always thought."

"It sounds like you all learned some pretty difficult lessons," Kurt said, supportively.

"One of the greatest regrets I had was the example I set for my son. In fact, by my careless I-can-do-anything attitude, I gave him *permission* to become reckless too.

"It took me a while to figure it all out, but the grief Edith experienced, Jason's problems and my own injuries could all have been prevented.

"It's like the what-if game ... what if I had used the gloves Edith offered ... what if she had removed the keys from the ignition ... or taken Jason into the house until I had finished what I was doing ... or what if I hadn't been so headstrong ... so bullet-proof?

"At any rate, after seeing the damage I caused my family and in my own life, I decided to make safety my personal crusade," Sam explained. "Incidents can be prevented. That's the message I want to get across to others. It takes someone willing to speak up and it also takes the other person being willing to listen."

Kurt reached across the table and gripped Sam's arm. "I never guessed how painful your lesson was or how strongly it impacted your family."

"Just remember," Sam said. "Having an injury-free workplace is an achievable goal because incidents *are* preventable. Safety happens by choice and so do incidents. That's why every individual makes a difference."

✦ ✦ ✦ ✦ ✦

Later that night, after Kurt and Jessica had gone to bed, Kurt had difficulty getting to sleep. He couldn't stop thinking about Sam's story.

"Honest feedback was given and someone chose to ignore it," he thought, envisioning Jason seeing his father's bloody hand. "It is a lesson I don't want anyone to have to go through."

Ten minutes later, Kurt was ready to close his eyes, but before falling asleep, he whispered a prayer of thanks – for Sam, his wisdom and his willingness to help someone who needed his guidance.

Organizational change is a long road ... and he knew he was just beginning his own journey. Minutes later, Kurt was sleeping.

SNAPSHOT

DEFINITIONS:

1. **Safety champions** – people who are leading the charge, encouraging safety conversations between all levels of staff, and ensuring behaviors we want to see are rewarded.

ESSENTIAL CONCEPTS:

✦ When we build a *Safety 24/7 Culture*, we're strengthening our existing safety culture by addressing the human element.

✦ The *Safety 24/7 Culture* is powerful enough to transcend geographic and cultural differences because human behavior is the same wherever we go.

✦ What we need to focus on are those things people have in common, no matter what country they are from.

✦ People like the *Safety 24/7 Culture* because it matches the desire every worker has to go home from work in the same shape he was in when he arrived.

✦ A person's underlying motivations for working safely are universal.

✦ The skills necessary to be a good safety leader are the same skills necessary to be a good operations leader and vice versa:

(1) Build a foundation of trust.
(2) Set clear goals and high expectations.
(3) Praise progress as people begin changing behaviors.
(4) Have the courage to keep going.

✦ Operations and safety are 'interdependent.'

✦ Safety is as much everyone's responsibility as it is for the people who work in the safety department.

✦ Having an injury free workplace is an achievable goal because incidents **are** preventable. It takes someone willing to speak up and it also takes a person's willingness to listen.

✦ Safety happens by choice and so do incidents. That's why every individual makes a difference.

EPILOGUE

It had been almost a year since Kurt Bradshaw's promotion to manager of worldwide safety, and the past year had raised the bar, not only for Kurt but also for the rest of the organization.

His transition from operations to taking on the challenge of strengthening the company's safety culture, had not been without its glitches. However, the strengths Kurt had developed as a leader over the years were the same skills he required to develop a culture of safety. With Sam's mentoring, Kurt had met management's goal of reversing the skyrocketing number of claims, not to mention the high cost of operational downtime associated with all those injuries and fatalities.

Comparing last quarter's frequency rates to those of the previous year, there was no doubt: the majority of the team had taken ownership of *Safety 24/7*.

The company had no fatalities during the 12-month period. Their Lost Time Incident Rate had decreased over 30 percent. And, while there was still an alarming number of Near Hits, something that had not been previously measured, Kurt was pleased to see they were successful in creating an environment where people were now willing to report an incident had taken place.

The company was leading instead of trailing the industry in terms of safety performance.

The phone rang; Kurt answered to hear Ron Kaiser's voice. "Seen the safety numbers for last quarter?"

"Just going over them now," Kurt replied. "They're continuing to improve – that's the good news."

"I can tell you, the executive management team and board of directors are nothing short of euphoric. One of our clients even told us our improved safety record was key to being awarded that large contract last month, and our new culture of safety was the hot topic during this quarter's analyst call," Kaiser was almost gushing. "Of course, we're not where we want to be, but as the president said during the safety workshop *he* attended, 'commitment to *Safety 24/7* will help us reach our goal of everyone making it safely home to their loved ones.'"

"He's right," Kurt agreed. "Of course, we're never going to eliminate all the risks, but if we continue to encourage observation, and reinforce the behaviors we want to see through *Safety 24/7* Conversations, we will set a new standard for the industry."

"Right … right on," Kaiser said, his voice beaming, "Just keep it up … good job, man!"

Kurt continued to scan the report. As of that quarter, all personnel – more than 4,000 in 20 countries – had completed a *Safety 24/7* workshop. But Kurt knew there was no 'silver bullet' when it came to training, which is why they were already developing follow-on workshops.

He sat back … proud of what the organization had achieved. The return on investment was clearly evident. The reduction in claims costs alone had more than paid for the development and implementation of *Safety 24/7*. Plus, the company was seeing the benefits of the operational improvements fostered by the new culture, and clients were awarding additional contracts based upon their improved safety record.

"Maybe more importantly though," Kurt thought to himself, "people are actually happier coming to work. They're 'proud to be a part of the team', as one supervisor had commented during a recent workshop."

Another had said, "It's an attitude – this safety culture. People take their responsibility seriously when it comes to keeping members of the team, as well as themselves, safe. I didn't think I'd ever say this, but safety has become part of everything I do both here and at home … I just see things in an entirely different light."

Just then the phone rang. "Mr. Bradshaw," said the voice at the other end. "Mr. Jacobsen has asked that you come to his office."

"Uh, right away," Kurt responded as he put the receiver in its cradle. Being summoned to the president's office had not happened very often in his career.

As he took the elevator and walked down the hall, Kurt prepared himself. Could it be another crisis in Saudi Arabia? Or South America? He was still guessing as he stood in the office doorway.

"Bradshaw … come in, come in." Fred Jacobsen welcomed him warmly, making sure Kurt knew the rest of the people seated in the room.

"Sorry for this spur-of-the-moment meeting, but we've made some decisions, one of which involves your area," Jacobsen continued. "We've all reviewed our safety results and want to tell you how impressed we are. The numbers have never been this good."

Kurt smiled, "Thank you, it's been a team effort."

"Agreed, and we're behind it 100 percent," Jacobsen said. "In fact, the reason we asked you to join us was to let you know we're elevating the importance of safety within the organization. We recognize truly committing to a culture of safety is integral to the success of our business."

"Thank you for your confidence in *Safety 24/7*," Kurt responded, genuinely pleased.

"And," Jacobson continued, "In order to continue demonstrating our commitment, we are promoting you to vice-president of health, safety and environment."

Kurt had not expected a promotion … not this soon. "Well," he began, "I'm surprised and flattered, needless to say …"

Just then, Kurt noticed Fred Jacobsen was frowning … and he was looking in Kurt's direction, making the newly-announced VP wonder if he had said the wrong thing. Then he saw the reason for Jacobsen's obvious displeasure.

The person seated next to Kurt had, out of habit, leaned his chair back and was balancing on the two back legs.

"Martin …," the president began, "As the senior Human Resources professional in our organization, you've helped the executive team understand the importance of leading by example. However, there is something that concerns me about your safety...."

Kurt smiled. Clearly, Fred Jacobson believed in *Safety 24/7* – moving safety from his mind to his heart, where real change takes place.

Multiple factors provide the framework for, and allow definition of, a behavior-based safety program. What follows are specific attributes of *24/7 Safety*, which enable it to make such a positive and significant difference within an organization:

1. Placing emphasis on human characteristics, rather than specific behaviors in the workplace, provides the understanding required for individuals to incorporate safety into their lives on a 24/7 basis.

2. Elevating safety from a priority to a core value demonstrates an unwavering commitment in the face of competing objectives. This enables the individual to have the courage necessary to think about his or her behaviors prior to an incident taking place.

3. Identification and elimination of at-risk behaviors addresses the foundation of incidents, while acknowledging risk itself is an inherent part of life.

4. Acknowledging similarities in *human* culture allows for geographical expansion of an organization's *safety* culture.

5. Recognition of the interdependency of safety within the workplace enables the individual to utilize leadership skills not unique to safety.

At the risk of sounding redundant, we hope you will use *Safety 24/7* in your own personal journey and, as a result, will work to improve safety continuously wherever life takes you.

Remember, real change can only take place where there is an open mind and a willing heart.

About the Authors

Gregory M. Anderson was formerly President of Intertek's Consulting & Training division, which specializes in providing behavior-based safety, leadership, teambuilding and intercultural diplomacy for organizations operating in high-risk environments. With offices around the world, the company has conducted workshops for over a quarter of a million people in more than 60 countries.

Greg, who is a graduate of the University of Southern California, has lived, worked and traveled in more than 40 countries as he battled oil fires in Kuwait, provided infrastructure for military personnel in Haiti and drilled for oil in North Africa. He currently provides independent consulting services while living in Seattle with his wife and three daughters.

Robert L. Lorber, Ph.D., is President and Chief Executive Officer of The Lorber Kamai Consulting Group. With extensive experience in the mining sector, Bob is focused on management effectiveness and has implemented productivity improvement systems at medium-size and Fortune 500 companies on five continents.

Co-author of The New York Times bestseller *Putting The One Minute Manager To Work*, with Kenneth Blanchard, Bob has also co-authored *One Page Management* with Riaz Khadem, as well as several other titles. He is an internationally-recognized expert on performance management, productivity, teamwork and the relationship between organizational structure and culture. Bob serves as an executive coach to numerous CEO's and executive teams.

Bob teaches as visiting faculty at the UC Davis Graduate School of Management and is the Chair of the Deans Advisory Council. He currently serves on the UC Davis School of Medicine's Board of Visitors and many other boards of corporations and not-for-profit organizations. He and his wife Sandy live with their three daughters in Davis, California.

ACKNOWLEDGMENTS
by Greg Anderson

I believe you can learn something from every person you meet, and I want to take this opportunity to thank those people who have provided me their knowledge.

The team at Intertek Consulting & Training for their hard work and uncompromising standards. You are the finest group of people I have ever had the pleasure of working with.

Randy and Helen Smith for entrusting me with so much.

The men and women of GlobalSantaFe. Your unwavering commitment to FOCUS not only made this book possible but, more importantly, it literally has changed people's lives for generations to come.

Bob Lorber for walking beside two generations of the Anderson clan.

Ken Blanchard, who has taught so many of us not only how to be better managers, but how to be better human beings.

Alice Adams, who can turn a blank canvas into a work of art.

Karen Stewart, who knows what I am thinking, usually before I do.

My parents, Gordon and Elizabeth Anderson, the two finest people on earth. Nothing I could write would express my love and appreciation for all you have given me.

My daughters, Brittany, Makenzie and Kendall, for understanding there is no book on how to be the perfect dad. I am so proud of the person each of you has become.

Finally, my best friend and wife, Robin. Thank you for believing and helping me become more than I ever thought possible. You are the foundation for all that is good in my life.

ACKNOWLEDGMENTS
by Bob Lorber

To the four special Ladies in my life – my wife Sandy, and my three daughters Tracie, Lindie and Kaylie. Thanks for putting up with me! I love you all very much.

The numerous consultants at The Lorber Kamai Consulting Group who worked so hard throughout the years improving safety and saving lives all over the world.

Gordon Anderson, mentor and friend, who made most of this all possible. Leadership, kindness, patience and teaching to everyone around him. Gordon, you made this all possible!

Neal Parry, Red Barnhill, and Alan Bernardo – teachers, friends, and clients who taught me about the Drilling Industry and how to work in the many cultures around the world.

Mark Rosen for your friendship and leadership in Safety Training for our company in coal mines across this country.

Ken Blanchard – mentor, teacher, friend and colleague for over 30 years. You help everyone around be a better person and helped me really understand the meaning of servant leadership.

Mick Ukleja for your friendship, unconditional support and guidance throughout these past 22 years.

Greg Anderson – You made this book happen, and it will make a difference in thousands of peoples lives. It is a pleasure working with you and becoming part of your family.

Bring a *Safety 24/7* Culture
to Your Organization

Are you, like Kurt, recognizing the need to reduce incidents and injuries within your organization? Is the cost of claims and lost productivity having significant impact on the bottom-line? Or maybe the organization has a good safety record but there is concern you may have reached a plateau and you want to keep the momentum going.

If so, we have the expertise to help you expand your culture of safety and demonstrate the improvement other organizations have achieved. Additional ways we can help instill the *Safety 24/7* principles within your organization include:

DETERMINING NEEDS
You have limited resources to invest on even the most critical issues; therefore, you want to be sure you're focused on the right ones. Are employees aware of their behavior and the impact it has on safety, or have they grown complacent over time? Are supervisors tolerating at-risk behaviors because of *perceived* pressure? Or is senior management unintentionally sending "mixed messages"? We can help you determine what actions will be most effective in expanding your organization's safety culture.

SENIOR MANAGEMENT PLANNING
Is senior management uncertain where, how … or if … to start improving your organization's safety performance? Our staff of experienced professionals can design and facilitate an executive team

meeting to help clarify the organization's approach to creating a company-wide safety culture and the results that can realistically be accomplished.

WORKSHOPS

To achieve zero incidents, employees at all levels of the organization need to receive a consistent message on their role in building a strong safety culture and learn skills for turning those responsibilities into action. Our curriculum designers and facilitators will work with you to develop and conduct workshops to address your specific objectives while celebrating the many positives of your organization.

ON-SITE COACHING

True behavior change requires on-going feedback. We can work with your managers' onsite to develop the skills they need to create a work environment where employees are motivated to want to work safely.

Please visit our website,
www.safety247.org,
or call us at 832-654-9099 to discuss how
we can help you bring *Safety 24/7* alive in your culture.

Order Form

1-30 copies $16.50   31-100 copies $15.50   101+ copies $14.50

Safety 24/7   ＿＿＿＿copies X ＿＿＿＿ = $ ＿＿＿＿

English is standard.

Please specify if you prefer:

     __Spanish   __French   __Portuguese   __German

     __Arabic   __Chinese   __Swedish   __Japanese

Shipping & Handling Charges $ ＿＿＿＿

Sales Tax 10.74% (Washington Only) $ ＿＿＿＿

Total (U.S. Dollars Only) $ ＿＿＿＿

*Excludes Sales Tax – Except in Washington

Domestic U.S. (AK, HI add'l) Shipping & Handling Charges
(Single point delivery. 'Handling Charge' fee only, w/use of company shipping account.)
International Quoted on Individual Order Basis - See Contact Details Below

Shipping and Handling Charges

Total $ Amount	Up to $50	$51-$170	$171-$685	$686-$1165	$1166-$2999	$3000+
Charge	$7	$16	$30	$50	$80	$150

Name＿＿＿＿＿＿＿＿＿＿＿＿＿＿＿＿＿＿＿ Job Title ＿＿＿＿＿＿＿＿

Organization＿＿＿＿＿＿＿＿＿＿＿＿＿ Phone ＿＿＿＿＿＿＿＿＿

Email ＿＿＿＿＿＿＿＿＿＿＿＿＿＿＿ Fax ＿＿＿＿＿＿＿＿＿＿

Shipping Address ＿＿＿＿＿＿＿＿＿＿＿＿＿＿＿＿＿

City＿＿＿＿＿＿＿＿＿＿＿＿＿ State ＿＿＿＿ ZIP ＿＿＿＿＿＿

Billing Address & Contact ＿＿＿＿＿＿＿＿＿＿＿＿＿＿
(if different than shipping)

Purchase Order Number (if applicable)＿＿＿＿＿＿＿＿＿＿＿

Charge Your Order: ❑ MasterCard   ❑ Visa   ❑ American Express

Credit Card Number ＿＿＿＿＿＿＿＿＿＿＿ Exp. Date ＿＿＿＿＿＿＿

Signature ＿＿＿＿＿＿＿＿＿＿＿＿＿＿＿＿＿＿＿

Phone: 832.654.9099

Website: www.safety247.org

Mail orders to:
RGA Holdings, LLC
PO Box 99823
Seattle, WA 98139-0823
USA